Playing Along

THE OXFORD MUSIC /MEDIA SERIES

Daniel Goldmark, Series Editor

PLAYING ALONG
Digital Games, YouTube, and Virtual Performance

Kiri Miller

OXFORD
UNIVERSITY PRESS

OXFORD
UNIVERSITY PRESS

Oxford University Press, Inc., publishes works that further
Oxford University's objective of excellence
in research, scholarship, and education.

Oxford   New York
Auckland   Cape Town   Dar es Salaam   Hong Kong   Karachi
Kuala Lumpur   Madrid   Melbourne   Mexico City   Nairobi
New Delhi   Shanghai   Taipei   Toronto

With offices in
Argentina   Austria   Brazil   Chile   Czech Republic   France   Greece
Guatemala   Hungary   Italy   Japan   Poland   Portugal   Singapore
South Korea   Switzerland   Thailand   Turkey   Ukraine   Vietnam

Published by Oxford University Press, Inc.
198 Madison Avenue, New York, New York 10016

www.oup.com

Oxford is a registered trademark of Oxford University Press

Library of Congress Cataloging-in-Publication Data
Miller, Kiri.
Playing along : music, video games, and networked amateurs / Kiri Miller.
    p.   cm. — (Oxford music/media series)
Includes bibliographical references and index.
ISBN 978-0-19-975345-1 (alk. paper) — ISBN 978-0-19-975346-8 (alk. paper)
1. Video games. 2. Video games—Social aspects. 3. Interactive videos. 4. Video game music.
5. Popular music. I. Title.
GV1469.3.M55   2011
794.8—dc23   2011018803

1 3 5 7 9 8 6 4 2

Printed in the United States of America
on acid-free paper

"I am hooked on the charm of making the dumb machines sing."
—Janet Murray, *Hamlet on the Holodeck*, 1997

"The flesh of even the virtual performer remains too solid, and will not melt."
—Steve Dixon, *Digital Performance*, 2007

CONTENTS

ACKNOWLEDGMENTS

Like an episode of digital gameplay, this book is a collaborative performance. I can't hope to name all the individuals who have contributed their time, ideas, and support, but I can make a start at thanking those who have played leading roles. The reference list at the back of this volume should be considered a continuation of these credits.

First, my thanks to the players, practitioners, designers, entrepreneurs, and teachers at the heart of this book: Freddie Wong, Rob Kay, Shaun Scovil, Mike Dadmun, Brian Shandra, David Taub, Nate Brown, Nate Torres, John Dean, Grimmly, and all the volunteers who completed surveys, participated in gameplay/interview sessions, graciously replied to my follow-up emails, and commented on draft chapters. You made this book happen.

My Oxford University Press editor, Norm Hirschy, has been unfailingly generous and uncommonly discerning, and he has demonstrated sustained enthusiasm for this project over several years—pure manna to an author slogging through drafts. Thanks are also due to series editor Daniel Goldmark, production editor Erica Woods Tucker, copyeditor Elliot Simon, and the Oxford University Press editorial, production, and promotions staff for all their work on this project.

My two extraordinary undergraduate research assistants, Kate Reutershan and Emily Xie, made direct and substantive contributions to this book. I was very fortunate to have their help, which was made possible through the support of Brown's UTRA program and Radcliffe's Research Partnership Program.

Many mentors and colleagues in academia (including several dedicated journal editors) helped me refine my arguments and offered crucial encouragement along the way. They include Harry Berger, Katherine Bergeron, Bettina Brandl-Risi, Gabriele Brandstetter, Mark Butler, Tim Cooley, Kai van Eikels, Jessica Enevold, Dana Gooley, Tomie Hahn, Judith Hamera, David Josephson, Henry Klumpenhouwer, Leta Miller, Marc Perlman, Regula Qureshi, Butch Rovan, Simone Pereira de Sá, Kay Kaufman Shelemay, Jason Stanyek, Michael Steinberg, Jonathan Sterne, Rose Subotnik, Jeff Titon, Kay Warren, Todd Winkler, and several anonymous peer reviewers.

Thanks to all of my departmental colleagues at Brown—particularly Jim Baker and Shep Shapiro during their terms as department chair—for offering the personal and administrative support that made it possible for me to bring this project to completion. The peerless staff of the Brown Music Department helped me keep my head above water in my first years on the tenure track. My thanks to Jen Vieira, Kathleen Nelson, Mary Rego, and Ashley Lundh. In the Orwig Music Library, Ned Quist, Sheila Hogg, and Nancy Jakubowski offered fantastic research support, sending me a steady stream of scholarly and popular references that I could never have tracked down on my own. I am also indebted to the students in Musical Youth Cultures, Music and Technoculture, and Ethnography of Popular Music for helping me think through the core issues in this book over the years—and particularly to Colin Fitzpatrick and Liam McGranahan, whose thesis and dissertation projects on technoculture topics taught me a great deal. Rameen and Sarah Peyrow, Dan Boyne, and Jill Manning, my ashtanga teachers, offered me a complementary education in sensational knowledge that informed every page of this book.

Many friends have sustained me and offered invaluable insights over the course of this project. They include (roughly in order of appearance): Jesse Kurlancheek, Molly Kovel, Lilith Wood, Megan Jennings, DJ Hatfield, Te-Yi Lee, Carolyn Deacy, Aaron Girard, Christina Linklater, Myke Cuthbert, Mary Greitzer, Victoria Widican, Natalie Kirschstein, Petra Gelbart, Anneka Lenssen, Sindhu Revuluri, Jon Bernhardt, Kevin McCoy, Sheryl Kaskowitz, Ben Shaykin, Cindy Boucher, Niyati Dhokai, Jessica Keyes, Heather Hutchinson, Vanessa Ryan, Paja Faudree, Sandy Zipp, Peter Torelli, Joshua Tucker, Jessa Leinaweaver, Betsey Biggs, and Katherine Mitchell.

I finished this book during a dreamlike year at the Radcliffe Institute for Advanced Study. My thanks to Judith Vichniac, Sharon Lim-Hing, Melissa Synott, Marlon Cummings, and all my brilliant colleagues in Byerly for making this such a productive and inspiring research leave. I am also very grateful for the material support I received from a Killam Memorial Postdoctoral Fellowship at the University of Alberta in 2005–07, a Strothman Faculty Research Award at Brown in 2009–10, and a fellowship from the American Council of Learned Societies in 2010–11.

Thanks as always to the many branches of my family (virtual and actual). My partner, James Baumgartner, has been the perfect muse for this book. A gamer, DJ, cellist, radio producer, bike blogger, and master parodist, he has influenced my work more than he knows.

This book is dedicated to the next generation of players, including Ash, Caroline, Dahlia, Darwin, Dezh, Ezra, Julia, Leo, Micah, Nyo, and Patrik.

Portions of *Playing Along* were first published in the following articles:

"Grove Street Grimm: *Grand Theft Auto* and Digital Folklore." *Journal of American Folklore* 121(481): 255–285, Summer 2008. Copyright © 2008 by the Board of Trustees of the University of Illinois. Reprinted with permission of the University of Illinois Press.

"Jacking the Dial: Radio, Race, and Place in *Grand Theft Auto*." *Ethnomusicology* 51(3): 402–438, Fall 2007. Reprinted with the permission of the Society for Ethnomusicology.

"The Accidental Carjack: Ethnography, Gameworld Tourism, and *Grand Theft Auto*." *Game Studies* 8(1): n.p. (http://gamestudies.org/0801/articles/miller), Sept. 2008. Copyright © 2008 *Game Studies*.

"Schizophonic Performance: *Guitar Hero*, *Rock Band*, and Virtual Virtuosity." *Journal of the Society for American Music* 3(4): 395–429, Fall 2009. Copyright © 2009 The Society for American Music. Reprinted with the permission of Cambridge University Press.

ABOUT THE COMPANION WEBSITE

www.oup.com/us/playingalong

Oxford has created a password-protected website to accompany *Playing Along*. It includes audio recordings of interviews with players, excerpts from many of the online videos discussed in the book, and links to those videos in their original context. Following these links will allow readers to experience not just the videos but their associated comment threads, which play a key role in the book.

While the interviews with players are also transcribed in the main text, readers are particularly encouraged to listen to the interview recordings, since players' voices convey nuances that are lost in transcription. In some cases the audio clips also include more material than is quoted in the book.

Multimedia examples available online are found throughout the text and are signaled with ▶.

Playing Along

Introduction: Playing Along

ARE YOU READY TO PLAY?

You're grasping a controller designed to fit your hands. Gently nudging twin joysticks with your thumbs, you weave through L.A. traffic on your way to your next mission. It's one you've tried and failed before: sneaking through a rich rapper's mansion to steal his rhyme notebook, which might help boost the career of a friend who's trying to make a name for himself in the music industry. You scan through the car's radio dial and settle on the classic soul station, Master Sounds—it's funky and relaxing. But when Charles Wright's "Express Yourself" ends, a station ID reproves you, anticipating your larceny and your friend's plagiarism: "When you're done stealing beats, gangstas, listen to Master Sounds." You grin and share the joke with the game designers, wherever they are. Then you pull the car over, enter the rapper's mansion, and focus on your mission—peering around corners with your avatar's eyes and listening for the guard's footfalls with his ears, the controller in your hands mastered and forgotten.

You're grasping a controller designed to fit your hands—well, almost. There are five buttons on the neck of the guitar, but you only have four fingers available to press them. You keep your left index finger tensed, ready to change positions for the chain of hammer-on riffs that you can see coming inexorably down the screen. Adrenaline makes your face flush; you hold your breath as you play the end of the familiar chorus and start the solo, working out a fingering pattern that avoids your weak pinkie finger whenever possible. Your drummer whistles appreciatively; her part is easy for the moment, so she's watching your notation track. "Nailed it!" you say—and in your relief, you flub the first notes of the new verse.

You're grasping an instrument that feels designed to remake your hands. Your fingertips burn as you press them into the strings; your recalcitrant knuckle joints refuse to pivot as you reach to form a chord. You remind yourself that David says the first real callus is "a merit badge," something to look forward to. You stare into the screen as he exhorts you to strum along with him for two minutes, just straight eighth-notes on an easy chord. You're confident in your ability to keep a beat, but still you can't keep up—fingers hurt, wrist hurts, pick slips, strings sound dead. David can't see you quitting and restarting in the middle, so you feel guilty when he congratulates you at the end. You slide the scroll bar back and play the video again. If you can just get through that two minutes once, you'll give yourself permission to try the next lesson.

You're pressing your palms and feet into a sticky mat designed to frame and stabilize your practice; your body looks like an inverted V. This might be your millionth downward-dog. You listen to your breathing, focus your eyes, fractionally tilt your pelvis, try to be weightless—maybe focus on pulling up instead of pressing down? The length of mat between your hands and feet is a looming chasm. You think again about the slow-motion jump-through video you've just watched and try to make that body your own: jump your feet off the mat, hover for a moment with your weight transferred to your hands, keep your arms long and straight as you pivot at your shoulder joints, pull your legs through the space between your hands, and land on your behind, legs straight in front of you and hands now pressing into the mat next to your hips. Impossible. Halfway through the jump, you catch your toes on the mat and land awkwardly cross-legged, for the millionth time. You get up and watch the YouTube video again, search for some screenshots, reconfirm that other humans can do this. You find a blogger in the U.K., Grimmly, and browse his 57 posts on the jump-through. He writes, "No matter how many times you watch a video or look at these pictures, when you actually come to do it there's still this mental block, there's something counterintuitive to the whole thing, the floor is too damned close." He tells you to try it with your eyes closed: "Jump to your arms and then bring your head up which kind of pivots you through." You do it. Your feet don't touch the floor.

This book is about forms of play, performance, and embodied practice that bridge the virtual and the visceral. I focus on digital media and participatory culture: video games, online discussion forums, YouTube videos, blogs, and the communities of practice that they gather. The four portraits of "playing along" that began this introduction highlight a central theme: that both digital media and embodied knowledge can bridge space and time, creating connections between dispersed and diverse individual human experiences. The coming chapters show how digital media are brought to bear in the transmission of embodied knowledge: how a *Grand Theft Auto* player uses a virtual radio to hear with her avatar's ears; how a *Guitar Hero* player channels the experience of a live rock performer; how a beginning guitar student translates a two-dimensional prerecorded online video lesson into three-dimensional physical practice and an intimate relationship with a distant teacher; how a yoga practitioner relearns the possibilities of his own body by participating in the "cybershala."

I have brought together these practices under the heading of "playing along" because this phrase draws attention to fundamental intersections among gameplay, musical experience, and theatrical performance. Playing games, playing music, and playing a role onstage are all forms of collaborative performance, framed by rules and repertoires, structured through repetition (Kapchan 1995:479). Particular performance techniques create "relational infrastructure . . . by rendering bodies readable, and by organizing the relationships in which these readings can occur" (Hamera 2007:19). These cultural practices are so ubiquitous that they are often pressed into service as models for actual or desired human social relations: society as drama, as game, as making music or dancing together (e.g., Turner 1974, Geertz 1980, Schechner 1985, Desmond 1993–1994, Titon 1997, Sutton-Smith 1997, Small 1998). When scholars address the experiential *specificity* of games, music, or theater—the traits that make them compelling as practices rather than as metaphors—their arguments often center on immersive engagement. All these forms of "playing" have the capacity to "absorb the player intensely and utterly," structuring participants' sensory experience and altering their sense of the passage of time (Huizinga 2003 [1938]:45; see also Schutz 1964–67 [1951], de Marinis 1987, Csikszentmihalyi 1991, Grau 2003).

Playing *along* reminds us of the collaborative nature of these practices: that they build on (and build up) relationships among game designers, players, choreographers, dancers, writers, composers, directors, performers, and audiences, as well as marketers, publishers, and other commercial mediators. For instance, *Grand Theft Auto* is primarily a single-player game, but no one ever plays it alone; each player collaborates with the game designers to turn code into virtual performance, while remaining aware that millions of other players have engaged in the same endeavor. Common experiences of this kind generate "sensational knowledge" that connects dispersed participants, be they video game players, rock drummers, or yoga practitioners (Hahn 2007).

The "along" also raises questions about individual agency. In everyday social relations and in theater, "playing along" implies self-aware complicity: making a social contract, suspending disbelief, or actively creating belief (Murray 1997:110). The moral panics stirred by violent games like *Grand Theft Auto* hinge on an assumption that players are not *just* playing along, but are instead internalizing their avatar's criminal inclinations. Meanwhile, in the realm of music, "playing along" implies collectivity with an undercurrent of conformity: merging with a group or accompanying a recording, as opposed to being a creative individual or a virtuosic soloist. *Guitar Hero* and *Rock Band* players are often assumed to be "just playing along" with prerecorded tracks—an assessment that figures them as dupes of the mass-culture industry, deluded by fantasies of rock stardom.

These complex connotations of "playing along" suggest particular analytical pathways through the tangled thicket of contemporary participatory culture (Jenkins 2006b). Who calls the tune or makes the rules of the game? How do people learn to play, and why do they choose to join in? All the practices discussed in this book are voluntary leisure activities, and they require investments of time and money. How do they generate the affective experience that justifies those investments—that is, what makes them engaging, fun, satisfying, and meaningful? If "playing along" is inherently conformist, might it also leave some space for creativity? If it requires complicity, what are the terms of its social contracts? If it is always collaborative but increasingly mediated, how are collaborators connected across space and time?

By bringing together these diverse case studies, I aim to illuminate playing along as a "genre of participation" that transcends particular media formats and industry genre distinctions (Ito, Baumer et al. 2010b:18). The pursuits discussed here are massively popular. Recent surveys have found that 67 percent of American households play video games and 58 percent play musical instruments, rates that vary only slightly when broken down by gender, race, education level, or household income. While discussions of the positive or deleterious effects of such activities tend to focus on children and teens, 65 percent of instrumentalists and 75 percent of game players are over 18.[1] All of these practices are part of the same media ecology, one in which "listener," "gamer," and "musician" are not mutually exclusive categories (Horst, Herr-Stephenson et al. 2010). In the course of my research I have encountered individuals who model the whole trajectory of this book: They first heard a song while playing *Grand Theft Auto*, then actively used that song to create the right atmosphere for a particular gameplay episode, then bought the song as downloadable content for *Rock Band*, and eventually sought out online video tutorials for playing the song on guitar or drums. These practices all play along with established popular music canons and are constrained by cross-media licensing agreements. Commercial forces shape participants' practice across all the media domains under discussion—but without wholly dictating the nature of their experience.

"Playing along" also has methodological implications: It evokes participant-observation and ethnographic fieldwork. This might seem obvious. How could one possibly understand participatory culture without participating? But ethnographic studies of popular music and digital media remain outnumbered by analyses that treat all media products as "texts," suitable for close reading according to the interpretive conventions of literary theory. Digital-game scholars are somewhat more attuned to interactivity, practice, and process, but their arguments are often driven by rules-oriented analysis that assumes gameplay is sealed off from players' everyday lives—as though the constraints of game rules and the immersive qualities of virtual play could completely override one's prior knowledge, life experience, and critical faculties.[2] A performance-oriented approach grounded in ethnographic methodology brings different perspectives to this burgeoning field of cultural production and criticism: that is, *the perspectives of players and producers*, as well as the analytical insights they inspire (cf. Willis 1990). This book does offer detailed interpretation of media texts, but always in the service of understanding actual media practices.[3]

Popular media ethnographers face particular challenges. With millions of media consumers scattered around the world, how can we make meaningful claims about common experiences or values? The ethnographic "field" for this kind of work is hard to locate or circumscribe. Its research sites include a preponderance of Internet-based communities, leading some to wonder whether digital ethnographers are interacting with flimsily constructed virtual personae rather than real people enmeshed in genuine social relationships.[4] Industry research poses its own challenges. Even if I manage to make contact with someone inside a company, he or she may not be allowed to talk on the record. (At the beginning of one interview with a game designer, I handed him an interview consent form and he handed me a nondisclosure agreement.)

Many ethnographers of popular culture now point to in-depth work in particular locales as the literal grounds for their interpretive conclusions. As Martin Stokes writes, "The top-down perspective has been localized: Specific global projects are understood in specific institutional contexts" (2004:50). In a sense I will do the same here: I address particular games, websites, repertoires, and musical practices as frameworks for cultivating shared experiences and common knowledge. My fieldwork sites are "affinity spaces," where "people relate to each other primarily in terms of common interests, endeavors, goals, or practices, not primarily in terms of race, gender, age, disability, or social class. These latter variables are backgrounded, though they can be used (or not) strategically by people if and when they choose" (Gee 2004:85). This is not to say that questions of identity are irrelevant to the practices addressed in this book; to the contrary, racial representation and gendered performance are key themes. However, rather than undertaking deep fieldwork in a single localized community and exploring how individuals negotiate and perform identities within that community, I have structured my inquiry around the "common interests, endeavors, goals, or practices" that define affinity spaces: play, performance, experimentation, interpretation, teaching, and learning. These practices encourage participants to experience a sense of connection *across* differences—not necessarily "imagined community," but mutual understanding and common ground.

In addressing "virtual" contexts for play, performance, and pedagogy, I follow anthropologist Tom Boellstorff's lead in assuming that "our 'real' lives have been 'virtual' all along. It is in being virtual that we are human: since it is human 'nature' to experience life through the prism of culture, human being has always been virtual being" (2008:5). In his ethnography of the virtual culture of Second Life, Boellstorff argues that there is nothing unreal about virtual experience; rather than opposing virtuality and reality, he follows Aristotle in distinguishing the virtual from "the actual." As Marie-Laure Ryan puts it, "In scholastic philosophy 'actual' and 'virtual' exist in a dialectical relation rather than in one of radical opposition: the virtual is not that which is deprived of existence, but that which possesses the potential, or force of developing into actual existence" (1999:88). Thus virtuality exists "whenever there is a perceived gap between experience and 'the actual.' . . . 'Virtual' connotes approaching the actual *without arriving there*" (Boellstorff 2008:19). The case studies in this book represent a spectrum of practices that take place in that in-between terrain between virtual and actual experience; they each "approach the actual" in a different manner and to a different extent. In fact, in some cases they arrive at actuality—as when an online guitar student or a cybershala yoga practitioner makes the jump from imaginatively channeling a technique to performing it in her own body.

Theorists of virtuality too often assume that our actual bodies are irrelevant to contemporary techno-mediated life. As Steve Dixon notes, many digital-performance scholars push theories of "the metamorphosis and fragmentation of the body in virtual realms" so far that they end up believing in disembodiment (2007:215)—whereas "what is *practiced* by digital/posthuman performers is commonly the search for the opposite: for cohesion, for meaning, for unity, for intimate cybernetic connections between the organic and the technological" (155). In a related line of discourse, some scholars of massively multiplayer online worlds draw hard lines between synthetic, designed, virtual domains and the real, physical world—"the world of earth, air, fire, water, and blood that we've inherited from our forebears," as Edward Castronova puts it (2007:7). In his book *Exodus to the Virtual World*, Castronova predicts that "the desire to be 'in there' as opposed to 'out here' will strengthen" (58), and further asserts, "This is a competition. People compare the real and the virtual, and then they decide" (69). This approach precludes the possibility of being between "in there" and "out here": playing at the interface of virtual and visceral experience. In this book I argue that we should be more attentive to players' "in-between" embodied experiences and interpretive practices: their ability to move in and out of immersion at will and to let their virtual and actual domains of experience inform one another.

In addressing these "in-between" practices, I draw on the methodological and theoretical orientations of ethnomusicology, performance studies, popular music studies, and a growing body of interdisciplinary digital-media scholarship. I bring a different toolkit of approaches to each case study discussed in the book, but all of my work builds on the assumption that "playing along" with practitioners is the best way to understand their complex, diverse experiences with new digital media.

GRAND THEFT AUTO: PLAYING THROUGH CJ'S STORY, HEARING WITH CJ'S EARS

A *Gamezilla* reviewer summed up the *Grand Theft Auto* series this way: "I know what it is, you know what it is, and more likely than not, even your mom knows what it is. In our community of gamers, it's known for the immersive and innovative gameplay that gives the player freedom unlike any other game out there. In the mainstream media and everywhere else, it's known as that one game that teaches kids how to hijack cars and pick up hookers" (Lucas 2004). While lawyers have reaped substantial paychecks from lawsuits related to the games' scandalous content (Glater 2008), GTA's sandbox-style structural "freedom" has been acclaimed by players, industry critics, and game theorists as the most appealing and innovative feature of the series.

The GTA gameworlds are based on major American cities: New York City becomes Liberty City in *GTA III* (2001) and *GTA IV* (2008), Miami becomes Vice City in the eponymous game (2002), and Los Angeles, San Francisco, and Las Vegas become Los Santos, San Fierro, and Las Venturas in *GTA: San Andreas* (2004). Players have no choice of avatar; in each game they take on the persona of a well-defined underdog, most recently a poor black gang member under pressure from corrupt police (CJ, in *San Andreas*) and a fresh-off-the-boat Serbian immigrant with an ugly war history behind him (Niko Bellic, in *GTA IV*). Completing missions gradually raises the lead character's status and expands his territory, but players are also free to suspend goal-oriented, plot-advancing activities and simply explore the gameworld. Each virtual city comes equipped with a variety of vehicles and weapons, stores and restaurants, and local media (including in-game radio, TV, Internet, and parodic advertising to suit each of these formats).

These games seem tailor-made for narrative-oriented analysis. They have well-defined protagonists who live in specific American cities at particular historical moments (e.g., *Vice City* is set in 1986 Miami; *San Andreas* focuses on 1992 Los Angeles). Each game presents a large-scale plot arc built up from episodic "missions." Narrative analysis can certainly generate insights into story-driven games, revealing resemblances across many genres of creative production (Murray 1997). For instance, elsewhere I have compared *GTA: San Andreas* to the Grimm brothers' compendium of children's tales, "a stylistically and ideologically standardized storybook [that] reinforced earlier narratives and influenced the formulation and the maintenance of the tales in both oral and literary circulation" (Dégh 1979:84; see Miller 2008a). But if we move beyond the notion of the game as storybook and the game designers as writers/compilers, other aspects of the gameplay experience become available for investigation.

How is playing through a narrative different from reading it, recounting it, or hearing it told? In *Cybertext*, an important early contribution to digital-game theory, Espen Aarseth characterized the difference between conventional linear texts and "cybertexts" in terms of individual agency:

> When you read from a cybertext, you are constantly reminded of inaccessible strategies and paths not taken, voices not heard. Each decision will make some parts of the text more, and others less, accessible; and you may never know the exact results of your choices, that is, exactly what you missed. . . . Trying to know a cybertext is an investment of personal improvisation that can result in either intimacy or failure. (1997:3–4)

Aarseth's work on "ergodic literature"—texts which require this sort of improvisatory investment—helped shape Gonzalo Frasca's equally influential distinction between literary-theory-oriented "narratology" and game-theory-oriented "ludology" as the two major scholarly approaches to digital games (Frasca 1999; cf. Juul 2005:15). Ludologists concern themselves with the unique properties of games as distinct from narrative texts, including their variable outcomes and the effort they require from players. Juul's definition foregrounds the qualities important to ludologists: "A game is a rule-based system with a variable and quantifiable outcome, where different outcomes are assigned different values, the player exerts effort in order to influence the outcome, the player feels emotionally attached to the outcome, and the consequences of the activity are negotiable" (2005:36).

This game model has great potential as a frame for ethnographic work, but most ludologists—like most narratologists—have been more engaged with theory than with performance practice. Digital-game theorists frequently invoke Huizinga's notion of a "magic circle" that creates a sealed frame around gameplay. For example, "A game of Tetris . . . provides a formalized boundary regarding play: The game is either in play or it is not" (Salen and Zimmerman 2004:95, cf. Juul 2005:164–65; following Huizinga 1955).[5] By contrast, scholars who work on traditional children's games often focus on the porousness of that magic circle, the ways in which players muddy the distinction of being "either in play or not." As Brian Sutton-Smith has observed, "We treat play too often as a separable text, when in fact it always exists complexly interacting with the various contexts—human and symbolic—of which it is a part." Sutton-Smith urges scholars to consider "the years of effort, practice, preparation, organization, and the multiplicity of learned moments that have gone into this production" (1995:283; cf. Schechner 1985:16). I follow the lead of these scholars in foregrounding players' experiences (including my own), rather than engaging in formalist analysis of a game's narrative or its rule-based structure.

For the ethnographically inclined, the ludologists' emphasis on the improvisatory negotiation of unique paths through an established text or a rule-based system might suggest connections to theories of oral transmission, ritual, and traditional games—all of which involve individual improvisation within specific constraints. Anthropologists, folklorists, and ethnomusicologists routinely decry the notion that anonymous tradition-bearers just go through the motions required to replicate a given cultural text; their work offers reminders that video-game players aren't automatons, either. For instance, we might use oral-formulaic theory to model players' creative work as they string together episodic missions into different performances of what always counts as "the same" game (cf. Murray 1997:194). Each episode of digital gameplay might be viewed as what Jeff Titon calls a folkloric text, which always "exists in multiple versions and variants, similar to one another and thereby referencing one another," resulting in an expressive form with "an emergent, processual character" (1995:439; see also Motz 1998:348). *GTA: San Andreas* is a mass-produced text—a piece of computer software—yet satisfying gameplay relies on the unique realization of its code through individual performance.

In Part One of this book, I show how players cycle between immersion and critical detachment as they play along with GTA's guiding narratives, encoded repertoires of embodied performance, and ironic meta-commentary. In Chapter One, I work through several analytical frameworks for *GTA: San Andreas*, addressing the game as a tourist destination, a fieldwork site, a virtual museum, a vehicle for vicarious embodied performance, and a pop culture artifact whose double-voiced aesthetic has given rise to diverse interpretive communities. I argue that the GTA games encourage players to adopt touristic, ethnographic, and colonialist orientations to gameworld exploration. These games are fundamentally about controlling territory: Each lead character must acquire authority over threatening and disorienting spaces. Players act out and internalize this narrative theme; the tricks and secrets of the terrain must be learned, named, and mastered (Mechling 1984). Players also come to understand the specific cultural logic of each gameworld through firsthand experience. As they play, their strategic experimentation and fortuitous blunders highlight the gap between their own physical abilities, learned behaviors, and life history and those of their avatars. *GTA: San Andreas* offers a particularly rich case study in virtual performance because its lead character is African American (an extreme rarity in video games). In *San Andreas*, playing along means performing in blackface.

In Chapter Two, I focus on musical life in *San Andreas*: how players use the game's radio stations and music industry stories to understand the gameworld and its characters. For instance, many players choose to "hear with CJ's ears," tuning the radio dial to the hip-hop stations because that's the music their avatar would prefer. New listeners as well as longtime hip-hop heads have reported that listening to West Coast gangsta rap while playing *San Andreas* shaped their interpretation of both the music and the game. As one critic wrote, "Although we've been listening to NWA's anthem, 'F*ck the Police,' since we sprouted chest hair, we only now appreciate the pathos that suffuses the lyrics" (Naqvi 2005). While the GTA series dwells outside the "music/rhythm" game-industry niche, these games make music an integral part of everyday life in the gameworld—as it is in the actual world.

GUITAR HERO AND ROCK BAND: SCHIZOPHONIC PERFORMANCE

In the early twentieth century, technological innovations in the mechanical reproduction of music began to reach a large market of middle-class consumers. The perceived affordances of the player piano, the phonograph, and the radio delineated a new distinction between live and recorded performance.[6] These technologies also inspired a century of eulogies for amateur musicality. Mark Katz notes that early critics of the phonograph worried that it would "deter amateur performance and turn Americans into passive musical consumers"; by 1916, an organ teacher had reported that some students "gave up study when they secured a mechanical machine, as it enabled them to enjoy correct performances of music which they could never be able to execute" (Katz 2004:68). Two decades later, recorded music was ubiquitous; Theodor Adorno asserted that contemporary performance "sounds like its own phonograph record" (2001 [1938]:44) and argued that "the adaptation to machine music necessarily implies a renunciation of one's own human feelings and at the same time a fetishism of the machine" (1941:41). In 1969 the Canadian composer R. Murray Schafer coined the term "schizophonia" to refer to "splitting the sound from the makers of the sound" (1969:46). Around the same time, Roland Barthes mourned the disappearance of "musica practica," his term for music experienced through the act of amateur performance (1977 [1970]). More recently, Jacques Attali characterized mass-produced recorded music as "a means of silencing, a concrete example of commodities speaking in place of people" (1985 [1977]:111).

Every advance in sound-reproduction technology seemingly dealt another blow to amateur musical practice. By the mid-1990s, Michael Chanan had dubbed the Walkman "the final coup in the negation that recording perpetuates on musica practica, where instead of music coming from bodies in front of the listener, it is reduced to an unreal and intangible space enveloping the isolated head" (1995:155). In a book charting the history of musica practica, Chanan argued more broadly that "the widespread practical knowledge of music which comes from its educated amateur practice has been displaced and demoted. It is part of this syndrome that the integral place occupied by music in liberal humanist education has been lost, and music has become an increasingly isolated technical study" (1994:8).

But one could also recount a parallel version of this history, a version in which hybrid forms of "mediated and live" musicality proliferate around every corner (Keil 1984). Ever since it became possible to reproduce recorded performances in domestic settings, people have been engaging in musical practices that fall between the poles of passive listening (if there is such a thing) and musica practica. In the early days of commercial recording, music publishers supplied "piano reductions of works commonly available on disc so that music lovers could perform their favorite recorded music themselves" (Katz 2004:70). Player pianos often included mechanical features that invited consumers to control tempo, accent, sustain, and other elements of playback (Taylor 2007:285). In 1916, "Oscar Saenger published a course of vocal study in which the student listened to and then imitated various exercises on several specially made discs" (Katz 2004:70). We should also remember that listening can be deeply participatory; consider the art-music connoisseurs praised by Adorno, who grasp "musical sense" by "spontaneously linking the known elements—a reaction as spontaneous by the listener as it was spontaneous by the composer—in order to experience the inherent novelty of the composition" (1941:33). Late in his career Adorno noted that "The ability to repeat long-playing records, as well as parts of them, fosters a familiarity which is hardly afforded by the ritual of performance," and he vividly described the process by which "recordings awaken to a second life in the wondrous dialogue with the lonely and perceptive listeners" (1990 [1969]:64, 66; cf. Frith 1996:203–204 on popular music). Moving from participatory listening toward musica practica, we might consider various forms of mimetic performance (air conducting, air guitar, lip-syncing); playing along with recordings (including the music-minus-one recordings produced specifically for this purpose); and learning new repertoire or techniques by ear from recorded examples.

Charles Keil, an ethnomusicologist known for his work on the social and musical dynamics of live ensemble performance, apparently had something of an epiphany when he experienced "a very different cultural adaptation to electrically mediated music" on a visit to Japan. After encountering karaoke, he found himself wondering, "Why haven't I paid any ethnomusicological attention to the fact that since I was a teenager I have spent more time playing bass or drums to records than I have playing with other people?" (1984:92, 95). As Keil notes, such practices are easy to overlook, especially when they take place in private, domestic settings. They are also difficult to categorize, since they have often emerged in cultural contexts where people make hard distinctions between musicians and nonmusicians, live performance and recordings, playing and listening, and, more generally, original creation and reception/consumption.

Since the 1970s, an assortment of popular music genres and practices have increasingly eroded, blurred, and undermined these distinctions (Goodwin 1992, Théberge 1997). Jamaican dub engineers pioneered studio remix techniques, and the "sound system" collectives who brought those recordings to a listening, dancing public developed new performance styles that built on the particular affordances of available playback technology (Veal 2007). In New York, some of these same practices contributed to the development of hip-hop (Rose 1994a). With the development of karaoke in Japan, "music-minus-one" amateur musicianship became a public, participatory performance genre that has since spread around the world (Mitsui and Hosokawa 1998, Lum 1996, Drew 2001). Turntablists, electronic dance music producers, and laptop musicians have challenged assumptions about what counts as an instrument, a live performance, or an original composition (Schloss 2004, Butler 2006). Meanwhile, the copy/paste/remix/redistribute capacities of digital-media technologies have reshaped the relationship between commercial production and consumption (Messaris and Humphreys 2006, Jenkins 2006a, Sinnreich 2007, McGranahan 2010).

Nevertheless, while definitions of live musical performance and individual creative authorship sometimes stretch to encompass new techno-mediated practices, they have also retained their longstanding connotations of authenticity, uniqueness, and value—in part because of the crucial role these ideologies play in generating revenue for the star-driven commercial music industry. Liveness and individual creativity remain particularly important in the world of rock music, where live performance "certifies" the authenticity of recordings and visual artifacts; rock listeners "are tolerant of studio manipulation only to the extent that they know or believe that the resulting sound can be reproduced on stage by the same performers" (Auslander 1999:76, 82).

This thumbnail sketch of the intersecting histories of sound reproduction technology, amateur musicianship, and commercial popular music sets the stage for *Guitar Hero* and *Rock Band*, video games that promise players they can be rock stars in their living rooms (see Figure I.1). These games index both a revival of *musica practica* and a host of anxieties about inauthentic musicality and mindless consumerism. *Guitar Hero* players follow color-coded notation that streams down their TV screens, indicating which plastic buttons to press on a toy guitar. If they hit the right buttons, someone else's classic recording (or a faithful cover version) comes out of the speakers. It's hard to imagine a greater affront to rock authenticity. Indeed, as I show in Chapter Three, many people think it's self-evident that the *Guitar Hero* and *Rock Band* games are creating musical automatons who suffer from escapist delusions of rock stardom. Critics of the games often argue that if there is a saving grace here, it can only reside in the possibility that the scales will fall from players' eyes and they'll be inspired to pick up real instruments.

The *Guitar Hero* and *Rock Band* franchises sold over 46 million game units and brought in over 3 billion dollars between 2005 and 2010.[7] At the end of 2009, *Guitar Hero III* ranked second in a list of the all-time top-selling console games released in the United States (behind *Wii Play*, which came bundled with the Nintendo Wii console, and just ahead of *Grand Theft Auto: San Andreas*; see Thorsen 2010). In an era of mp3 file-sharing and declining album sales, these music-oriented games created a huge market for value-added versions of previously recorded popular music. Every song licensed for release in

Figure I.1
"Rock Band," xkcd comic #359 (xkcd n.d.).

the games has been broken down into parts and transcribed at four different difficulty levels, creating a new, hard-to-pirate digital music product. Once players have bought a game and a set of instrument controllers and have invested the time required to achieve proficiency on one or more instruments, they are happy to spend money on new repertoire. This is a venerable business model, one that previously drove sales of four-hands piano transcriptions, sheet music for parlor ensembles, player-piano rolls, and subsequent recorded music formats (Christensen 1999, Taylor 2007).

The song transcriptions, game scoring mechanism, and on-screen avatar musicians are the most obvious "value-added" components of *Guitar Hero* and *Rock Band*; this is the design and coding work for which game company employees are paid. But I argue that the most important value-added aspect of these games is the potential for performance. Actually, the term *value-reconstituted* might be more appropriate: you reconstitute instant soup by adding water, and you reconstitute a recorded song by adding performance. *Guitar Hero* and *Rock Band* let players put the performance back into recorded music, reanimating it with their physical engagement and adrenaline. Players become live performers of prerecorded songs, a phenomenon that I refer to as "schizophonic performance."

The instant-gratification aspect of schizophonic performance is distasteful to some people, especially those who are invested in "the dedicated artistry and drive for individualistic self-expression embodied in the figure of the guitar hero" (Waksman 2001:119). Instant fame is for industry-manufactured sellouts, and hitting buttons on a plastic controller to release someone else's hot guitar solo seems a lot like lip-syncing; it's not even as authentic as karaoke. But players aren't deluded. They're quick to point out that they understand the difference between playing instruments and playing *Guitar Hero*—and many of them do both. They know that the "instant" songs that they play in these games are packaged, commercialized, and designed to be labor-saving, but that doesn't spoil their musical experience. Just add performance, and the music blooms into new life.

In Part Two of this book, I argue that these games have inspired new modes of engagement with popular music. Tens of millions of players now use instrument-shaped controllers to play classic and contemporary rock songs. When they do so, they also encounter and assess game designers' and critics' conceptions of rock's canonical repertoire, aesthetic norms, performance conventions, and symbolic value. While many people have described *Guitar Hero* gameplay as "just playing along" with prerecorded material, I suggest that it is more accurate to think of it as playing *between*—that is, playing in the gap between virtual and actual performance. Chapter Three addresses *Guitar Hero* and *Rock Band* gameplay as a developing genre of collaborative, participatory rock music performance that is generating debate about the nature of musical and performative authenticity. I analyze the games' implicit models of rock ideology, their sometimes-sincere/sometimes-ironic constructions of rock heroism, and heated media debates about whether the games foster genuine musicianship. I also explore players' ideas about musicality and creativity, focusing on players who have significant experience playing traditional instruments. In Chapter Four, I turn to public gameplay performance contexts. These games inspire physically virtuosic, visually engaging performances, staged in a variety of social contexts: Players compete in game tournaments, post virtuosic performances on YouTube, and form instant bands with friends at *Rock Band* bar nights. Drawing on participant-observation, interviews with players and game designers, a web-based qualitative survey, and media reception analysis, in these chapters I investigate what it means to be a live performer of a prerecorded song.

ONLINE LEARNING AND COMMUNITIES OF PRACTICE

I first stumbled upon the world of online music pedagogy through my *Guitar Hero* gameplay. The games introduced me to the "hammer-on" guitar technique, which is presented as a special gameplay mode in which one can play rapid passages of notes on the guitar controller's fret buttons without using the strum bar. On a whim, I searched for "hammer-on" on YouTube, hoping to find out how this technique worked on an instrument with strings instead of buttons. I found a wealth of guitar lesson videos, interspersed with player-produced *Guitar Hero* lessons that taught fingering patterns for difficult hammer-on passages in the games.

YouTube.com—a free, public, online video archive with built-in social networking features—has created a platform for countless virtual communities, many of which are focused on transmitting knowledge in users' areas of interest and expertise. Some of these learning communities are gradually transforming the face-to-face, body-to-body transmission contexts that have always played a crucial role in music pedagogy. Classical and popular music and dance instructors, bedroom DJs, and masters of traditional musics from around the world are all engaging in these new forms of transmission, gathering committed students who view video lessons and post their own performance efforts online for community feedback. Amateur musicians are also posting and studying footage of musical performances from decades past (e.g., by classical piano virtuosos) in order to teach themselves performance techniques.

YouTube is not the only online platform for such activities, but it is the currently dominant streaming-video site—in part because it is a subsidiary of Google, the currently dominant internet search engine company. Thus YouTube often serves as a gateway. Even when people have the skill and inclination to design and host their own websites, they maintain an active presence on YouTube in order to reach new viewers and attract traffic elsewhere. As Burgess and Green note, YouTube functions as an institution, "operating as a coordinating mechanism between individual and collective creativity and meaning production; and as a mediator between various competing industry-oriented discourses and ideologies and various audience- or user-oriented ones" (2009:37).

YouTube is part of the broader "Web 2.0" phenomenon, characterized by business models and software development that support and capitalize on user-generated content, peer-to-peer interaction, and collaborative/participatory content production and editing.[8] These features highlight "the ability of digital networked media to create new publics and audiences for amateur work" (Lange and Ito 2010:284). Much academic and popular commentary about contemporary digital culture has focused on the production and circulation of "co-creative" media and amateur creative expression: blogs, citizen journalism, fan fiction, original or remixed music and videos, and so forth (e.g., Morris 2004; Jenkins 2006b; Taylor, T.D. 2006; Sinnreich 2007; Potts, Hartley et al. 2008). But online participatory culture is not just about circulating new texts, democratizing creative expression, performing one's identity in a networked public sphere, or spreading the latest news—nor is it about propagating "an on-demand, personalized version [of the truth] that reflects our own individual myopia," as Andrew Keen would have it (2007:17). Web 2.0 platforms provide new channels for teaching and learning, for transmitting practical knowledge and drawing together communities of practice.

Like digital games, online pedagogy engages participants in bridging the gap between virtual and visceral experience. In Part Three of the book, I show

how people are using streaming video, blogs, and online social networking to teach and learn actual-world practices: playing acoustic instruments and practicing yoga. In Chapter Five, I focus on two musicians who have used YouTube and their own websites to develop multimedia online curricula for teaching rock guitar and drums. David Taub's NextLevelGuitar.com and Nate Brown's OnlineDrummer.com have each attracted thousands of students from around the world. I discuss how Taub, Brown, and their students have strategically adapted traditional pedagogical strategies and reshaped teacher–student (and student–student) relationships in a virtual transmission context. This chapter also addresses online teachers' encounters with copyright law. Aspiring guitarists and drummers often seek out lessons because they want to learn to play songs by their favorite artists. Musicians routinely teach copyrighted material in private lessons, but publishing song lessons online can leave them vulnerable to account deletion or legal action. While bringing lessons into the digital public sphere makes them accessible to many more students, this new context also imposes new constraints on repertoire and teaching style.

Chapter Six shifts the focus to "amateur-to-amateur" online pedagogy. Here I present and compare three shorter case studies: a conga lesson series on YouTube created by a Virginia college student of Puerto Rican descent; YouTube-based piano lessons that cater to beginners who want to learn to play songs by the Beatles and Billy Joel; and ashtanga yoga transmission in a "cybershala" built out of online videos and blog posts. Each sheds light on how people are using interactive digital media to teach and learn embodied practices with established histories of actual-world pedagogy and performance.

The case studies discussed in this book represent a spectrum of virtual performance styles and forms of mediated musicking (Small 1998). In *Grand Theft Auto*, players engage with popular music canons and historical narratives in the course of exploring an immersive gameworld, but they have tightly constrained musical performance options (limited to controlling in-game radio stations). In *Guitar Hero* and *Rock Band*, embodied musicianship is foregrounded but detached from the actual production of sound. In online music pedagogy, teachers try to make their lessons as fun and engaging as video games while also seeking out ways to translate the face-to-face, body-to-body interactions of traditional private lessons for a web-based, public format. One aim of this book is to show what we might learn if we recognize that these practices rub shoulders, rather than ghettoizing them by genre, format, or commercial niche.

As a card-carrying ethnomusicologist, I am particularly concerned with breaking the artificial walls that have often separated music scholarship from other fields of inquiry—the walls that allow the authors of a magisterial volume on "digital performance" to note in passing that they won't be dealing with music, because "we do not have sufficient specialist knowledge and expertise to approach a worthy analysis," even as they acknowledge that "music was one of the first artistic fields to experiment significantly with and embrace computer technologies" (Dixon 2007:x). Instead of signing on to fill that gap—bringing to bear the "specialist knowledge and expertise" housed in a music department on the fringe of campus and addressing an audience that shares those guild skills—I want to suggest that both theorists and users of interactive digital media are already more musical than they might realize. Conversely, I hope that people who come to this book looking for research on video game music or online music lessons will not skip over the sections about virtual tourism, playing in an avatar's body, or practicing yoga with a laptop computer at one's elbow. The inclusive rubric of "playing along" is meant to invite readers to recognize the deep connections among these forms of play, performance, and pedagogy.

Technoculture is changing fast, and the specific games, websites, and media formats I discuss here are already becoming part of history. *Playing Along* documents some of that history, presenting detailed analysis of popular digital media practice in the early twenty-first century. I hope it will also raise broad questions with implications for newly emerging practices, illuminating your own everyday world of digital media, musical experience, and participatory culture.

PART ONE

Playing Along with Grand Theft Auto

CHAPTER 1

Straight Outta Ganton

Virtual Tourism, Fieldwork, and Performance

A little girl asks, "Uncle Ricky? Would you read us a bedtime story, please, oh please?" He replies, "All right, you kids get to bed, I'll get the storybook. Y'all tucked in? . . . Once upon a time, not long ago, when people wore pajamas and lived life slow. . . ."

The story he tells is a fable of urban violence and bad choices, about "a little boy who was misled" and fell into a life of crime. An action-packed play-by-play relates the climactic episode in this young man's life: he robs an undercover cop, escapes by pulling a gun, runs around the block, knocks over an old man in his haste, dashes into an abandoned building, gets another gun from a filthy drug addict, heads outside to steal a car, crashes it into a tree, escapes from the wreckage, takes a pregnant woman hostage but lets her go, runs again but is soon surrounded and shot by the police. The moral? "This ain't funny, so don't you dare laugh /Just another case about the wrong path / Straight and narrow or your soul gets cast. Good night!"

Slick Rick's hip-hop cautionary tale is titled "Children's Story" (Walters 1988). His laid-back, matter-of-fact delivery runs over a jumpy, menacing keyboard loop, the melody lifted from the bass line of Bob James's "Nautilus" (1974). Smooth jazz becomes sharp counterpoint, an itchy, edgy trigger finger. The subject matter is not unusual for late 1980s rap, but as it comes over the radio I'm startled into attention because Uncle Ricky seems to be telling my own story. Just moments ago, I foolishly robbed a pedestrian in front of a police officer, had to run "top speed til [I] was out of breath," accidentally knocked over a passing old lady, sought cover in an abandoned building to get my bearings, emerged to steal the nearest car, smashed it up and had to jump from the wreck, and finally found myself surrounded by police gunfire.

After I was flattened by the hail of bullets, however, I survived to find myself outside a hospital (with cash diminished and weapons gone). I carjacked some yuppie so I could get home, and it is his car's radio that is playing "Children's Story." As I drive, Slick Rick recounts the story that I have just performed, locating that performance within a particular narrative tradition: that of the violent, engaging, and often disturbing world of gangsta rap, whose storytellers trade in exaggerated portraits of urban street life coupled with trenchant critiques of contemporary society (Quinn 2005).

Once upon a time, not long ago, a company called Rockstar Games created a video game series called *Grand Theft Auto,* and 87 million people sat down to play (Take-Two Interactive 2008, Take-Two Interactive 2010). The GTA franchise now comprises six major multiplatform releases, plus about a dozen downloadable expansion packs and spin-offs for portable gaming systems (Moby Games 2010). Most of the GTA games are rated M (for Mature/17+) by the Entertainment Software Rating Board for their violence, language, and sexual content; *San Andreas*'s rating was changed to AO (Adults Only) when a hidden sex scene was discovered in the game after its release. Nevertheless, a 2010 study found that 56 percent of American 8- to 18-year-olds had played *Grand Theft Auto* (Roberts, Foehr et al. 2010:26).

My opening fieldwork story is drawn from my experience with *Grand Theft Auto: San Andreas* (Rockstar Games 2004a; see Figure 1.1). As the much-anticipated final element of the trilogy that began with *GTA III* and *Vice City*, *San Andreas* was a tremendous commercial success. It sold over 12 million copies in its first four months on the market. In the United Kingdom, it was the fastest-ever million-selling videogame—meaning that about one in sixty people in the U.K. bought this game within nine days of its release (Surette 2004, Adams 2005). These figures are not unusual when it comes to popular music or movies, but games like *San Andreas* are less casual purchases. They cost three or four times as much as new CD releases, a price point justified by the buyer's expectation of spending something like a hundred hours in the gameworld. Rockstar's collected tales of early 1990s gangsta life have been received and performed on a grand scale, worldwide, by players who visit CJ's world for hours at a time after dinner, before school, or between errands.

San Andreas is a fictional American state, modeled on the West Coast of the United States in terms of geography, architecture, climate, demographics, and popular culture.[1] My fieldworker's claim to direct experience of this terrain is complicated by the fact that it wasn't really me robbing pedestrians and jacking cars; it was my avatar, Carl Johnson, a young African American man whom I was maneuvering through the mean streets of Los Santos (the game's version of Los Angeles). Carl, who is nicknamed CJ, grew up on Grove Street in Ganton, a poor black neighborhood that stands in for L.A.'s Compton. *San Andreas* was released in 2004, but the gameworld is set in 1992—a watershed year for Los Angeles. That spring a jury acquitted the four police officers

Figure 1.1
Driving in the Vinewood/Hollywood hills. Promotional screenshot from *GTA: San Andreas* (Rockstar Games 2004d).

involved in the videotaped beating of Rodney King, sparking riots. Similar events transpire late in the game's storyline, when a corrupt police officer gets off easy and Los Santos goes up in flames. Named for California's major geological fault, *San Andreas* dwells on faultlines in American culture.

San Andreas's narrative is driven by CJ's personal circumstances, both material and emotional. The player first meets him in the Liberty City (New York City) airport; he has lived on the East Coast for five years but must return home to Grove Street for the funeral of his mother, who was murdered under mysterious circumstances. When this prodigal son arrives in Los Santos, he is greeted by city police, who rough him up, steal his money, let him know they are framing him for the murder of another police officer, and throw him out of their squad car to fend for himself in rival gang territory. This is when you first gain control of CJ's movements. Your first task is to get him onto a BMX bike and laboriously pedal through a disorienting, threatening urban landscape to reach his childhood home in Ganton/Compton. From there, CJ reunites with siblings and friends and returns to life with the Grove Street Families gang— fighting rival gangs to expand territory, beating up drug dealers who are destroying the neighborhood, and sometimes doing the bidding of the Los Santos police, who have him in a tough spot. There is no strict order for completion of the missions that will advance this story. At any given moment, players can choose among multiple tasks: for example, rescuing a family member held captive by a rival gang, competing in a lowrider car race through the streets of Los Santos, stealing weapons from a National Guard armory, or sneaking through a rapper's mansion to steal his book of rhymes. Completing all of the missions requires about 60 to 100 hours of gameplay, depending on the player's skill level.

This flexible narrative structure is a hallmark of the *Grand Theft Auto* series. In Gonzalo Frasca's assessment (building on Caillois's terms), the GTA games combine elements of *ludus* and *paideia*—the defining qualities of goal-oriented games and flexible "play" environments, respectively (Roberts 1999, 2003). Jesper Juul further suggests that the GTA games straddle the line between "emergence" and "progression" games, where "emergence" refers to games with a small set of well-defined rules that combine in complex ways (e.g., chess) and "progression" refers to games with a fixed series of tasks that tend to generate a narrative structure (2005:82–83). The GTA games combine the satisfactions of these genres. You can set the missions aside and simply explore the impressive expanses of the gameworld: steal a vehicle, turn on the radio, and cruise along beautiful beachfront highways. Or you could buy a weapon from the local gun shop and shoot at random passers-by, to see how they'll react and what useful items they might be carrying—until the police take notice and you wind up arrested or injured.

This flexibility is reined in by one crucial constraint: each GTA game has only one lead character. In many video games, players can choose from an array of characters or custom-build an avatar from scratch, choosing gender, race, physical attributes, and even skill sets. This freedom to experiment with one's virtual identity has been at the heart of most ethnographic work on cyberspace and digital gameplay. Scholars have analyzed social relationships in massively multiplayer online games (MMOGs) and virtual worlds like Second Life, in which people create distinctive avatars and use them to interact with other participants (e.g., Bartle 1996, Morris 2004, Taylor, T.L. 2006, Boellstorff 2008, Golub 2010; cf. Fine 1983). But there is no such interaction in *GTA: San Andreas*, nor even a choice of persona. This is a single-player game, and each player must wear the shoes of a particular young black man who finds himself in a particularly trying situation. When I said the violent protagonist in my lurid opening story "wasn't really me," it might have seemed like an abdication of responsibility—after all, who made CJ commit those acts if not the player at the controls? But in an important sense it really *wasn't* me—not just because no real shots were fired, but because this gameworld and its narratives revolve around CJ's race, gender, acquired abilities, family background, and other personal circumstances. From CJ's neighborhood friends to the hostile Los Santos police, every character in *San Andreas* sees CJ as a young black man in gang colors.

In this chapter I address gameplay in San Andreas as a form of virtual performance in which players experiment with various approaches to "playing along." The GTA games present irony-laced portraits of real American urban spaces and historical moments; they are as much works of cultural commentary as they are games. As players work their way through episodic missions and explore the vast virtual space of the gameworld, they cycle between immersion and detachment. Sometimes they play like method actors, sometimes like tourists, sometimes like experimenters in a lab or participant-observers doing fieldwork. They often document these experiences and share them with other players, both face to face and through online social media. In 1974, Michel de Certeau wrote about popular-culture reception practices in a manner that seems eerily appropriate to contemporary digital gameplay: He described "a certain art of placing one's blows, a pleasure in getting around the rules of a constraining space" that develops as people navigate popular culture's "established forces and representations" (1984 [1974]:18). Decades later, the GTA games show how the moment-to-moment art of placing one's blows and mastering someone else's space can accumulate into a collection of collaborative performances and documentary texts about place, representation, agency, urban realpolitik, and ethical subjectivity.

The analysis that follows relies on ethnographic sources from "technoculture," including online discussion forums, YouTube videos of gameplay sessions, game reviews by amateur and professional critics, email correspondence, and responses to a survey that I posted online.[2] The survey focused on interactive aspects of *San Andreas*'s musical soundtrack (see Chapter Two); I asked respondents follow-up questions about other aspects of gameplay via email correspondence. In addition to my web-based fieldwork, I have undertaken participant-observation. I have played the GTA games alone in my living room, as most of my survey respondents reported doing; I have played it while friends watched and commented; I have watched others play; and I have conducted post-gameplay interviews. The survey responses were collected between 2005 and 2007. Face-to-face interviews and gameplay observation took place during the same period in Edmonton, Alberta, Canada.

In many respects, my fieldwork has been quite traditional. Describing virtual spaces as field sites is no longer unusual; while some ethnographers still maintain that only a prolonged, immersive, on-the-ground encounter with an unfamiliar culture constitutes genuine fieldwork, there are now so many people interacting in virtual contexts that an ethnographic approach to their doings has gained broad acceptance.[3] GTA fansites and discussion forums therefore present rich field materials. But what would it mean for a scholar to say she is undertaking ethnographic fieldwork in the state of San Andreas—not only in gamer culture, but in a single-player gameworld? In this chapter I argue that "fieldwork" is not just a fruitful metaphor for guiding scholarly analysis of virtual worlds; it actually characterizes players' activities. When players visit the state of San Andreas, they are participant-observers in a culture not of their making. Their explorations blend touristic, ethnographic, and imperialist approaches. They evaluate the staged performances and scenery created for their amusement, they investigate the boundaries and social norms of the gameworld, and they gradually master new territory. All the while, they engage in a collaborative performance project with the game designers and millions of other players, playing along with or sometimes disrupting the lead character's narrative trajectory.

While players have considerable improvisational freedom in their approach to the GTA games, many of their experiences have been programmed to be reproducible. At the moment of my writing or your reading, thousands of people could be hearing Slick Rick's "Children's Story" on the car radio as they race away from a series of events much like those he has recounted. This is an important difference from the "persistent worlds" of massively multiplayer online games or online social spaces like Second Life, where the virtual environment has an ongoing history and a shifting population (see Taylor, T.L. 2006 and Boellstorff 2008). Every player experiences the storyline of *San Andreas* from the beginning, knowing that millions of others have done the same, and each person who finishes the final mission knows that she or he is graduating into a club of committed players who have also spent long hours in the gameworld. In the process, players develop "a system of mutual relevances," which encourages in-group communication and the formation of player communities (Fine 1983:238). This is how any player earns "storytelling rights," a sense of communicative entitlement and "ownership of experience" (Shuman 1986:18). Hours of practicing the physical skills that are required for successful gameplay also contribute to dispersed comradeship. Imagine the sense of shared experience that we might derive from having watched the same movie; now compare that to what we might feel if we had each performed the lead role in the same hundred-hour play.

A TOURIST'S GUIDE TO *GRAND THEFT AUTO*

Each game in the *Grand Theft Auto* series invites the player to navigate a complicated criminal underworld, undertaking a series of missions in a vast, detailed gamespace based on a major American urban center (see Table 1.1). The game designers rely heavily on irony, appealing to players attuned to political parody and popular culture as well as educating those not already so attuned. As Rockstar Games Creative Vice President Dan Houser put it, "It's leveled at the broader weirdness of America and American consumerism and American action movies" (Bramwell 2004). Political satire is a guiding design principle in all the GTA games, manifested in over-the-top parodies of American commercial radio, news media, and advertisements. Rockstar's designers also cite films, TV shows, and music videos in creating their gameworlds, sometimes ironically and sometimes in an evident spirit of homage. Archetypal characters, plot devices, and elements of cinematography are all borrowed from prior media productions, with a special debt to *The Godfather*, *Scarface*, *Miami Vice*, commercial rap videos, and hip-hop culture more broadly.

Table 1.1 *GRAND THEFT AUTO* MAJOR MULTI-PLATFORM RELEASES

game (year released)	gamespace
GTA III (2001)	Liberty City = New York, 2001
	avatar: unnamed white small-time crook
Vice City (2002)	Vice City = Miami, 1986
	avatar: Tommy Vercetti, Italian American mafioso
San Andreas (2004)	San Andreas = California/Nevada, 1992
	Los Santos = Los Angeles
	San Fierro = San Francisco
	Las Venturas = Las Vegas
	avatar: CJ, African American gang member from Compton
GTA IV (2008)	Liberty City = New York, 2008
	avatar: Niko Bellic, illegal Serbian immigrant/war veteran

Game scholars have noted that irony has become a "no-lose gambit" for the game industry, "a 'have your cake and eat it too' strategy whose simultaneous affirmation/negation structure can give the appearance of social critique and retract it in the same moment, . . . allowing practitioners to feel safely above it all even as they sink more deeply in" (Kline, Dyer-Witheford et al. 2003:277). This kind of gameplay experience is a hallmark of the GTA series, in which players routinely engage in violent rampages while, say, listening to an in-game radio ad that skewers American military recruitment tactics ("Only in the military would a teenager be given responsibilities like driving a nuclear submarine, maneuvering a tank, or dropping high-ordinance explosives!").[4] A similar form of "distanced immersion" is a key aspect of many tourist experiences, both in person and in virtual/cinematic contexts. Tourists experience conflicting desires to be fully immersed but also in "a comfortable viewing position" (Strain 2003:27). This fact was not lost on GTA's designers, who explicitly encourage players to adopt a tourist's perspective—"a gaze trained for consumerism" and "an instrument of mastery" (Strain 2003:15–16).

All of the GTA game instruction booklets take the form of tourist guides. The *San Andreas* booklet begins:

> San Andreas is one of the country's most diverse visitor destinations, known for its huge size and incredible variety. Although justly celebrated for its 3 wildly different main cities each with its own style and attractions to offer—Los Santos with its celebrity and sprawling ghettos, San Fierro with its eclectic artist community, and Las Venturas with the glitz and glamour [British spelling *sic*] of casinos—it has so much more beyond that; such as mountains, ghost towns, dense forests, and hot, dry deserts. Take time to visit the whole state and support local businesses. (Rockstar Games 2004b:5)

GTA IV's guidebook steadily alternates between earnest-sounding boosterism and nod-and-wink cues:

> With more sights, shows, restaurants, bars, museums, shopping and borderline psychopaths than you can shake a shotgun at, there is no place quite like Liberty City. Where else can you get sworn at by a senior citizen, accosted by a crackhead, propositioned by a prostitute and strip-searched by a police officer all before breakfast? Discover the history, culture and diversity that make this booming metropolis the capital of the world, at least according to the over-caffeinated locals. (Rockstar Games 2008b:4)

Game reviewers often follow Rockstar's lead; for instance, a U.K. *Mirror* reviewer described Liberty City as "a living, breathing, sprawling metropolis based on New York that you'll explore like a gawping tourist on your first few goes" (Lynch 2008). But many reviewers also point to the tension between adopting an ironic/distanced tourist perspective and identifying with one's avatar. Across the series the game protagonists have had increasingly complex personal histories and constraining identities: *San Andreas*'s CJ is a member of an inner-city African American underclass, for instance, a social position quite different from that of most tourists (and most video game players).[5] Being a native of Los Santos/Los Angeles, CJ would never need a tourist guide to get around town. Niko Bellic, *GTA IV*'s lead character, is indeed new to Liberty City, but he does not occupy a tourist's privileged position; he is an illegal immigrant with few connections, destined to discover that "the reality is very different from the dream in a city that worships money and status, and is heaven for those who have them and a living nightmare for those who don't" (Rockstar Games 2008a, back cover text). CJ's San Andreas and Niko's Liberty City could each be characterized as what sociologist Dean MacCannell calls "a front region that is totally organized to look like a back region," like a historical theme park's living diorama of the servant's quarters (2004 [1973]:204). With the help of the avatar, player-tourists can pass as natives in gritty urban underworlds.

In considering players' visits to these reimagined American cities, it's important to remember that GTA's players live all over the world and that many high-level Rockstar Games employees are European (including the company's British founders). Rockstar recreates these American spaces by engaging players in the reenactment of quintessentially American stories: Players around the world collaborate with the game designers in performances that comment on American popular culture and politics. Non-American players evaluate the realism and ironic elements of GTA's cities by drawing on their experience with other media depictions of these places, as several of my survey respondents explained in follow-up email correspondence:

> I do like the different areas with a different ethnic feel, e.g., Chinatown. Although most large cities have this, I think it is something everyone associates with the U.S. . . . The characters voices seem to rely on movie stereotypes such as mafioso

types. I think that this is quite a good thing when it comes to drawing you in and making you feel at home. (Bob, Scotland, 31–35)

The [radio] stations add to the "Americanness" of the game, as some of the content, especially on the talk radio stations, fits in with the stereotypical American radio: loud, opinionated, and ignorant. (Lincoln, Canada, 26–30)

I've never been to the US, which I expect changes my reaction to the game [compared] to that of Americans, particularly those that actually live in the cities— pretty much my entire knowledge of Miami or California comes from fictional films or TV shows, so I think subconsciously I consider them almost as fictional as 'Vice City' or 'San Andreas.' I think the one element of the game that really struck me as 'American' and that I doubt would work for me if the game was set in another first-world country is the sense one gets of seeing the police as 'the enemy' after a while. Maybe it's just because so many crime and gang-related movies are set in the US, but I don't think if my character was meant to be, say, a Canadian or a Frenchman in Canada or France, the concept of the police being an enemy I had to keep an eye out [for] wouldn't work so well. (David, Australia, 21–25)

I have never been to Miami, which was clearly the city that Vice City was trying to capture, so I can't comment on the accuracy of the portrayal. However, it was also trying to capture the Miami in Miami Vice and perhaps in Scarface. . . . In fact, I would say that Vice City was not supposed to be Miami in reality, but Miami in the media, and I feel that Vice City has accurately recalled this mediated Miami. (Jesse, Canada, age 21–25)[6]

These accounts demonstrate the porousness of the GTA gameworlds: Each player brings in his or her own expectations, biases, and a wealth of related media experiences, all of which have an impact on gameplay.

Touristic desires to collect authentic experiences through travel and consumption are well documented (Evans-Pritchard 1987, Britton 2004 [1991], Strain 2003). Because of this preoccupation with authenticity, many tourists—as well as many fieldworkers—continually suffer from what Martin Stokes has described as "semiotic anxiety," the perpetual nagging question of "is this real, or is this just a show?" (1999:143). But the double-voiced, ironic quality of the GTA series constantly assures visitors that it's all just a show, freeing players to admire the painstaking verisimilitude of some aspects of the gameworld without holding it to an "authentic" standard in every domain. Dan Houser touched on these issues of authenticity, realism, and tourist subjectivity in an interview with *New York* magazine about *GTA IV*:

We had guys looking at Census data; this part of Queens should be more Chinese. The [pedestrians] can go up and speak to each other now, so we got them speaking Russian, Spanish, Chinese. . . . We try to get the essence of the place, not a photo-realistic, digital tourist guide. We wanted a kind of spiritual tourist guide that feels like New York, but a blown-out, larger-than-life version. . . . We're not at all aspiring to virtual reality—what we are aspiring to is what feels like you're living in your own world. (Hill and Houser 2008)

That goal of intimate, individual ownership of the gameworld is furthered by GTA's predominantly single-player design.[7] GTA players can explore incredibly detailed terrain with touristic anonymity, never subjecting themselves to other players' judgments about their behavior, skills, or appearance (except by voluntarily sharing stories with other tourists later). In the curiously private space of GTA's virtual public spheres, one can experiment with behaviors that would be considered antisocial or hostile in many massively multiplayer environments: Going on an off-mission shooting spree in San Andreas will not inconvenience other players' avatars.

A shooting spree also isn't typical tourist behavior, of course, and it begins to suggest the relevance of alternative exploratory/experimental metaphors for GTA gameplay. Frasca speaks for many players when he describes the pleasures of "using the [GTA] environment as a giant laboratory for experimentation, where I could test the system's boundaries and set my own creative goals" (2003). To me, Frasca's descriptive terms for this game system recall a particular social scientific model of culture, a kind of separate-but-equal relativistic model in which cultures are "fenced off as culture gardens or . . . as boundary-maintaining systems based on shared values," in Fabian's critical assessment (1983:47). This model has been roundly critiqued in the anthropological literature for 30 years, in part because of the boundaries it erects between the ethnographer's culture and the culture under study. Today it is ethically untenable to regard someone else's culture as a laboratory. But in some respects the "culture gardens" model seems quite appropriate to a consideration of the GTA gameworlds. The assumption that Fabian critiques in Lévi-Straussian fieldwork—"that one person's immersion in the concrete world of another culture accomplishes the scientific feat of reducing that concrete world to its most general and universal principles" (61)—might make more sense in the GTA context, in which immersive, experimental gameplay gradually reveals the driving design principles, game challenges, and limitations of a bounded virtual environment. So, after considering these virtual cities as tourist destinations, what might one learn from conducting fieldwork there?

A FIELD EXPEDITION TO SAN ANDREAS

We might begin this thought-experiment by considering the family resemblances and distinctions among gameworlds, tourist sites, and fieldwork sites. Gameworlds and tourist sites are deliberately designed for the delectation of visitors;

traditional fieldwork sites are not. And yet in the theoretical frameworks and mission-style challenges that ethnographers establish for themselves, it isn't hard to see the architecture of something like a gameworld solidifying out of the cultural mist. Of necessity, ethnographers set themselves boundaries (of duration, locale, topic, theoretical model, and ethical practice) that delineate the terrain of "the field." Unsympathetic locals, observers from other disciplines, and post-"reflexive turn" ethnographers have been known to suggest that many fieldworkers are, in effect, constructing their own tourist sites or intellectual playgrounds; this is an aspect of Fabian's "culture gardens" critique, in that the garden is arguably both cultivated and enjoyed by the anthropologist (di Leonardo 1998, Agawu 2003). It is now widely acknowledged that the ethnographer's subjectivity is the central organizing principle of "the field," a conceptual space as constructed as any digital gameworld—and certainly just as shaped by collegial collaboration, a history of design precedents, and attentiveness to current trends.

GTA gameplay effectively models many aspects of a traditional field experience, in part because of the basic affinities of fieldwork, tourism, and play. As Nelson Graburn writes, "While human play may lack the travel element of tourism, it shares the aspects of removal from the normal rules, of limited duration and unique social relationships, and of the feelings of immersion and intensity that Turner characterized as flow. Like tourism, games are rituals which both differ from and reinforce certain aspects of the structure and the values of everyday life" (1983:95). This description implicitly aligns play and tourism with fieldwork, which also shares all these qualities. By supplying players with opportunities for travel, culture shock, new societal norms to absorb, and ethical quandaries to negotiate, the GTA games mimic "the 'being there' of classic ethnographic authority" (Marcus 1999:97), as well as fieldwork's rite-of-passage quality. There are even special field technologies to master: One must learn how to control the avatar, interpret city maps, and document one's experience (e.g., in *San Andreas*, by using the in-game camera, saving one's progress, and consulting the statistics the game provides to track CJ's changing skills, physical capabilities, and accomplishments).

The trickier issue is the question of where the ethnographer-player can locate ethnographic subjects, research associates, or "informants." After all, this is not a game in which players interact with other real people in the guise of avatars. While I would argue that GTA should still be considered as fundamentally social as a massively multiplayer online game, much of that sociality is either imagined or deferred until the game is turned off. I will suggest three possible answers to the missing-people problem of GTA fieldwork, each of which requires adjustments to the classic fieldwork ideal of direct communication with other people in their native cultural setting.

First, we could focus on the fact that the lone-explorer experience of the gameworld is a fiction. No one is ever truly alone in San Andreas, just as no one ever reads the newspaper, watches television, or surfs the net alone; all these media serve to constitute imagined communities of other readers, consumers, players, or citizens (Anderson 1991; cf. Fish 1980, Hermes 2005, Jenkins 2006b). The other players are invisible but unforgettably present. This axiom—no one is alone in San Andreas—points the researcher back to all those fansites, player-built mods, and discussion forums, but also highlights the importance of participant-observation in the gameworld. To be an effective interpreter of player discourse and player-created materials, and to converse with players on mutually respectful terms, I must have really been to San Andreas. In this sense the established relationship between credible ethnography and direct experience—the requirement to present evidence that *I was really there*—remains unchanged.

As a second approach, we could recognize the game's diverse team of designers, writers, actors, and culture consultants as the true native citizens of San Andreas. The in-game characters are vehicles for their real voices and physical gestures. Their political, aesthetic, and ethical sensibilities permeate the terrain. These research associates speak to the ethnographer with varying degrees of indirection and mediating irony; they comment on everything from the quality of commercial radio to the relative availability of firearms and nutritious food in contemporary American life. A perky announcer on a *San Andreas* talk radio station reports, "Culturally, this country is flatlining! Now you know why." Disrupting gameworld immersion with irony, Rockstar's writers invite the player-ethnographer to join in criticizing the object of their parody: mainstream American commercial media. Since products like the GTA games themselves are frequently invoked in that media as emblems of a "flatlining," culturally bankrupt America, the player who laughs at the radio announcer is forging an alliance with GTA's makers.

The hallmark of this second kind of ethnographic encounter is complicity, that sense of affinity between scholar and subject described by George Marcus as rapport's "evil twin" (1999:88). In general, the Rockstar team are perfect exemplars of the ethnographic subjects invoked by Marcus, James Clifford, and their disciplinary cohort. To use Clifford's terms, these individuals "turn out to have their own 'ethnographic' proclivities and interesting histories of travel. Insider-outsiders, good translators and explicators, they've been around" (1997:19). GTA's designers are fieldworkers and cultural producers in their own right, media-savvy and self-aware. In fact, by most standards their representational agency greatly exceeds that of any academic ethnographer. Thus this approach should probably be regarded as a form of "studying up" (Nader 1972). It requires the ethnographer to cycle between in-game fieldwork and in-real-life efforts to communicate with the game designers. Since most of them are famous, very busy, and legally constrained to watch what they say about *Grand Theft Auto*, the ethnographer occupies a supplicant position, obliged to cultivate her subjects' interest and good will—or to draw inferences based on interactions with intermediaries, such as in-game characters.[8]

The third way to locate ethnographic subjects in these games is the simplest: Suspend your disbelief and treat the gameworlds like actual places with human inhabitants. If we look at *Grand Theft Auto* through this lens, we might conclude that Rockstar has produced a thought-provoking training simulator for imperialist ethnography. Set in 1992 and featuring GTA's only African American avatar, *San Andreas* presents a dated, bounded, and ethically controversial model of a particular culture. As such, it calls to mind an ethnographic approach that many scholars now disown as similarly dated, confining, and discomfiting. How many early fieldwork narratives described a primitive society frozen in an era distant from our own, accessible only to this lone intrepid ethnographer? There, too, the informants were often people of color who spoke peculiar dialects and appeared incapable of truly understanding the ethnographic enterprise; they could be threatening or cooperative, but they were never complicit. Their concerns were strictly local, and with the aid of imported technology and a helpful native assistant the ethnographer could achieve a god's-eye view of their cultural system. (In GTA, the player generally observes the avatar from slightly above and behind but can change the camera angle at will.) There was little chance that these informants would ever read the ethnographer's work or comprehend their own supporting role in the drama of human civilization; in terms of agency, they might be likened to nonplayable characters.

GTA's space-structuring mechanisms also have imperialist-ethnography overtones. With the conceptual advent of virtual reality came a similar terrain-based paradigm for understanding digital "new frontiers"; consider the web browsers named Explorer, Navigator, and Safari. As Fuller and Jenkins point out, there are striking parallels between the developing rhetoric of virtual exploration and that of Renaissance travel writing on the New World. They suggest that not-yet-possessed digital game territories "exist only in the abstract, as potential sites for narrative action, as locations that have not yet been colonized" (1994:66). This analysis was based on early Nintendo games with strictly hierarchical game levels, but it is just as characteristic of the GTA games, where expanding and controlling territory is a central narrative theme. Early in *San Andreas* gameplay, most of the gameworld map is greyed-out to indicate that it is inaccessible and irrelevant. As Shira Chess observes in a Foucauldian analysis of the game, "as the player becomes more disciplined and adept at the game, [he is] given more space. Space, therefore, becomes a means of both disciplining and controlling a player, as well as a system of rewarding his acumen" (Chess 2005:82).

Colonizers, tourists, and fieldworkers have long shared a preoccupation with exploring and mastering new territory. But CJ is an anticolonist; his starting position is the postindustrial ghetto, and his colonizing explorations are conducted under duress. His persona and backstory, indelibly marked by race and class, are the foundation of the player's experience of place in San

Andreas—particularly of safe versus threatening places. CJ has to come back to Grove Street in Ganton again and again, and in the course of those repetitions this broken-down, impoverished neighborhood rapidly comes to feel like a safe and comforting destination. In Ganton there is a house where CJ can rest, keep his belongings, and change clothes; there are young men in CJ's gang colors on the streets; there are friends and family members who offer CJ jobs.

After only a few exploratory trips around Los Santos, I began to feel a wash of relief whenever I brought CJ back to Grove Street. The highway overpass and chain-link fences, the check-cashing place, the cars with mismatched doors, and the increasing number of black faces signaled safety. For a young middle-class white woman like me, as for many *San Andreas* players, this was a novel way to experience a poor black urban landscape. It reshaped my sense of public space and public safety, including my response to the sight of police (who tend to harass CJ, with or without provocation). It added a visceral dimension to my intellectual understanding of racial profiling. Eric Gwinn, a self-identified "black gamer" who is a staff writer for the *Chicago Tribune*, anticipated this effect in a review of *San Andreas*: "Soon after the game begins, the police stop the cab you, as CJ, are traveling in to let you know they're watching you. Are you being stopped because you're black or because the cops want you to know that as a former gangbanger, you're being watched? The answer is the latter. But for some players, the episode might cause a twinge, a glimpse into the mind of a black urban male" (2004). Such experiences have a long lineage in blackface performance; as Eric Lott notes, early blackface stars "inaugurated an American tradition of class abdication through gendered cross-racial immersion which persists, in historically differentiated ways, to our own day" (1993:51).

Is this hint of insight simply a byproduct of virtual imperialist ethnography? Are players positioned as colonizers of CJ's terrain, exploring a virtual ghetto that was always already there but didn't matter until it could be mastered by a video game player?[9] In considering this question, it's important to remember that the GTA games are suffused with irony and that they constantly invoke players' understandings of real-world places, politics, historical events, and existing media texts. Playing GTA means playing with one's sense of the past, experiences with pop culture, and ideas about America. *San Andreas* combines nostalgic idealization of an earlier era, ironic political commentary, and gameplay that evokes children's "cruel play" traditions in its brutality and inversions of social norms (Sutton-Smith 1983:109). In its nostalgic mode, *San Andreas* exemplifies the archaizing process through which "the repudiated is transvalued as heritage" (Kirshenblatt-Gimblett 1998b:298): the game's designers researched and reproduced the vernacular speech, clothing, music, and expressive aesthetic of a particular urban youth culture at a particular historical moment. In its ironic mode, which parodies American consumerism and stereotypes of ghetto life by using representational strategies appropriated from hip-hop, *San Andreas* suggests that this material is not yet "safe for preservation" (ibid.) but instead retains the capacity to provoke, alarm, or be judged offensively inauthentic.

Instead of a fieldwork site, then, we might address *San Andreas* as the product of the game designers' own fieldwork: a museum of vernacular culture, where visitors engage with interactive exhibits that they know were created for their own pleasure and edification. Unlike most large-scale reproductions of ye olde times or life among the natives, GTA is patently, gleefully insincere. Rather than dealing in pious "folklorism," creating a "semblance of the nonadministered, the original, the spontaneous, the naturally evolved" (Bausinger 1990:152), GTA's designers leave players free to tack back and forth between viewing particular elements as authentic or artificial, sincere or double-voiced.

PLAYING IN THE SAN ANDREAS MUSEUM

What was urban American life like in 1992? Find out at the San Andreas Museum, where painstaking research and craftsmanship have gone into producing every detail of three West Coast cities trapped in time. Try on period clothes in a variety of styles—trained reenactors representing native inhabitants will comment on your look! Take any car you want and speed through the streets so that you can see the reconstructed architecture from the same vantage point as real residents. While you're cruising around town, listen to a dazzling variety of authentic musical genres on the radio, where the playlists have been curated and presented by such luminaries as George Clinton, Chuck D, Axl Rose, and Julio G—a real-life radio DJ who spun gangsta rap in California back in the day. Grab lunch at the Cluckin' Bell or Pizza Stack, where actors representing apathetic teens will serve you the huge portions of high-fat fast food that were the main dining option for poor inner-city folk. The people you encounter will speak in a variety of slang styles, but their words will be subtitled for your convenience. What's more, no one will know you're a tourist! You'll be stepping into the shoes of a San Andreas native, complete with family and friends. Discreet screen instructions will guide you in observing local customs, but feel free to behave however you want. (Even the police will treat you like a local, except you'll never have to stay in the pen for long.) Other museums tell you about the music, cuisine, clothes, dialects, and social systems of past cultures with wall placards and glassed-in exhibits of stuff you can't touch. Only San Andreas puts you right in the action.

GTA's designers were the curators of this virtual museum, and like most curators they were preoccupied both with reaching a certain public and with staying true to the spirit of the exhibited material. A video game might not seem to bear a heavy burden in terms of representational accuracy and educational value, but the subject matter and setting of *San Andreas* imposed constraints on the Rockstar staff. First, many players would be qualified to assess the authenticity of Rockstar's version of 1992 West Coast culture; the game's music, cars, clothes, and dialogue had to stand up to scrutiny. In meeting this standard, the game acquired instructional force for those players who were *not* already connoisseurs of this material—it became a repository of canonical styles, songs, and slang. Second, the fact that CJ and most other characters are African American or Chicano generated considerable pressure on the game designers to avoid the appearance of racist caricature. In the run-up to the game's release, industry writers often suggested that it might be a risky business for European game developers to take on "American ghetto life." The protagonists of the previous games were white; *GTA III's* avatar was barely a character, and *Vice City's* campy *Miami Vice/Scarface* homage starred a cartoonish mafia thug who was impossible to take seriously. By contrast, a poor black man involved in gang violence in a run-down neighborhood seemed a dubious candidate for GTA's comic treatment.

GTA's designers aimed to preempt this line of criticism by invoking ethnographic authority. Chief operating officer Terry Donovan explained Rockstar's research process in a 2005 interview with *Electronic Gaming Monthly*:

EGM: So how did a bunch of guys from Scotland research all of San Andreas' inner-city content?

TERRY DONOVAN: Research is a really important part of development. It is vital to get the style and feel of the time and the place right. The team from Rockstar North went on a long research trip to the West Coast and traveled around photographing everything and absorbing everything. We also have a really meticulous team of researchers based in New York who obsess over every detail, and this combined with working closely with people like [tattoo artist] Mister Cartoon, [rap photographer] Estevan Oriol, and [screenwriter] DJ Pooh to draw on their knowledge and experience of the West Coast at that time helped to really put as much detail and feeling for the era into the game. (Donovan, Houser et al. 2005:106)

Rockstar gained credibility with cosmopolitan, hip, and media-savvy consumers by conducting extensive fieldwork and employing prestigious native informants—one might say they positioned their work as "polyphonic" ethnography (Clifford 1988:53). Their field research was not undertaken as an impartial survey of West Coast physical environments but rather as a means of collecting iconic snapshots of particular cities (the Vegas strip, San Francisco's hills and bridges, L.A.'s extreme contrasts of rich and poor neighborhoods) that could then be recreated using an aesthetic derived from 1990s hip-hop videos, album covers, graffiti, and "ghetto" films. Likewise, in choosing their research associates, Rockstar's staff did not seek out a representative cross section of ordinary city residents but opted to consult with rappers, lowrider club members, and people involved in past representations of hip-hop and "the ghetto," all of whom have an investment in a particular vision of the places and era represented in *San Andreas*. The San Andreas Museum was not meant to represent all the complexities of real inner-city life; instead, its curators sought to demonstrate their fidelity to hip-hop culture, through their aesthetic choices as well as the names in the game's credits (see Chapter Two).[10]

The most impressive interactive exhibit in the San Andreas Museum is CJ himself. He is the posthuman descendant of the exhibited exotic Others at World's Fairs and the costumed reenactors at Colonial Williamsburg: a walking, talking representative of a culture that has been fixed in time (Kirshenblatt-Gimblett 1998a:34). Again, this would seem to be a risky representational enterprise. Imagine a museum of American urban culture where visitors must put on blackface and gang colors at the entrance gates. Rockstar's designers addressed this risk by shifting a share of the burden of representation onto players. While *Vice City*'s mafioso avatar has just a few special outfits for certain missions, in *San Andreas* players can buy new clothes for CJ in the gameworld's many retail stores—everything from leisure suits to fancy watches to era-appropriate sneaker styles in a variety of colors. Players can also visit barbers and tattoo parlors to choose from an array of hairstyles and body art. They have to take CJ to a gym to build up his muscle rating, which affects not just his appearance but his physical capacities. CJ will lose energy and eventually lose weight unless his player takes him to a restaurant to eat. Since only fast-food restaurants are available, he can also gain weight rapidly—or throw up if he eats too much. Players engage in multiple levels of identity work through their in-game consumption practices. They may make choices in accordance with their personal tastes, but they must also help their upwardly mobile avatar "dress the part" in a manner that advances his goals.

These design innovations have major consequences for the gameplay experience. Consider the difference between playing as a young black man with short hair, a slender build, jeans, and a tank top—CJ's default appearance—or as one sporting cornrows, dog tags, gang colors, and tattoos, to cite only two among myriad options. (See Figures 1.2, 1.3, and 1.4.) One *San Andreas* fansite forum featured a thread called "What does your CJ look like?" There were over a thousand responses, including many screenshots. Here is a sampling of the verbal descriptions:

> So far i havent done any physical work with him, so he is average build. I have given him an afro, and he is kitted out with a green shirt, grey jeans and dogtags.

> No pic, but my CJ is a beast. Got him in warm ups and a track shirt, green hat and shoes. Worked out enough that his muscle is maxed out, he punched a cop car's door off. . . . Looks like a real muscle freak, like he really knew what to do with about 10 years in prison.

> I was at my friends house today, and since I had basically NO MUSCLE (I starved myself) We decided to make fat albert. We usually had 7–10 BIG meals EACH

Figure 1.2
Standard-issue CJ. Promotional screenshot from *GTA: San Andreas* (Rockstar Games 2004c).

Figure 1.3
Player-produced screenshot from *GTA: San Andreas* (igrandtheftauto.com 2004).

day. In about 3–4 days he was 100% fat. The only thing red jacket we could find was in Didier Sachs clothing section. It didn't look much like it, but it was the closest (the only, actually, it had a long v-neck, though.) we gave him blue jeans and black (brown, maybe) shoes. The afro was too big, so we gave him the jheri curls. CJ's face ruined it, anyway, but from the back it look like fat albert.

Figure 1.4
Player-produced screenshot from *GTA: San Andreas* (igrandtheftauto.com 2004).

If it isn't evident from the picture, my CJ looks like a mix of a bouncer and a hitman. Kind of cliche, but hey, I like it.

I have three CJ modes, depending on whether I'm near a wardrobe and whether I want to change or not. All of them are low-fat, max muscle: My standard CJ is dressed like me: green t-shirt, jean shorts, sandals, sunglasses, and a backwards cap (ok, I wear a beret instead of a cap, but still). "Cat burglar" CJ has a goatee and wears a beret, joke mask, black jacket, black khakis, and black sneakers. Basically, he looks like a Gangsta Beatnik. "Out-on-the-town-check-my-pimpin'-threads" CJ, which I rarely use, has a black boater hat, green suit, black pants, black dress shoes, a dollar chain, silver crowex, and sunglasses (all from DS [Didier Sachs])

lol [laugh out loud] i keep my cj in just heart boxers and thats it he goes out almost naked everywer
(igrandtheftauto.com 2004)

These players demonstrate a range of approaches to the aesthetic and ethical challenge of controlling CJ's appearance—and, more generally, of trying on a different body (Lahti 2003:167). Some have pushed the gangbanger stereotype to its logical conclusion: a "beast" who looks like he is fresh from prison (though the game's introduction tells us CJ has spent the last five years living in Liberty City).

Others push the other way, extending the game's ubiquitous intertextuality by making a comical "Fat Albert" from available materials. Still others dress CJ appropriately for different activities, with the default being "dressed like me."

These practices resemble those found in "fan art" communities. For example, Henry Jenkins has described how the production and circulation of *Star Trek*–inspired art "creates the conditions for a communal artform, one contrasting with the commercial culture from which it is derived in its refusal to make a profit and its desire to share its products with others who will value them" (1992:249). Software "mods" for games are created and circulated in a similar fashion. But *San Andreas* requires that every player—not just the hardcore fan or the skilled programmer—undertake these creative modifications of an established character, and that they do so in the course of everyday life in the gameworld. Sam Houser, Rockstar's president, explained the intended impact of this feature in an interview before the game's release: "By altering Carl's appearance, you are beginning to exist in his world, both in an obvious way—you are controlling how your character looks—but also in a less obvious way—all of your actions feel like they have consequences, and you are always in the world" (2004, cf. Lahti 2003:165). This curatorial decision not only changed the players' relationship to their avatar but relieved Rockstar of some representational responsibility. If players choose to dress CJ as a stereotypical gangbanger, pimp, or postprison "beast," that's their own problem.

For me, choosing a look for CJ did indeed feel like a problem. When I was first trying out hairstyle options at the barber, I had a visceral reaction to seeing CJ in an Afro: I felt guilty for turning him into a stereotype and immediately reverted to his standard-issue short hair. Like some of the forum contributors, for quite a while I dressed CJ only in heart-covered boxers. The incongruity of carrying out violent missions in this outfit lightened the mood of the game and saved me from feeling like I was outfitting CJ to be a racist caricature. But of course CJ *is* a caricature, by design and by definition. He can only become less of a caricature through a player's interpretive intervention. People who are uncomfortable playing along with an apparent stereotype have two main options for relief. They can trivialize CJ—as I did for a time with the heart-covered boxers, or as others do by creating an over-the-top gangsta parody—or they can make a complex pact with him: taking him seriously, identifying with him, and justifying his motives. Eventually I felt I was embarrassing CJ by making him look ridiculous. I took the advice of other characters in the game and got him some clothes in his gang's color, Grove Street green.[11]

THE ACCIDENTAL CARJACK: PLAYING IN CJ'S BODY

All digital game avatars are encoded with a repertoire of competences that players explore and exploit. Performing *San Andreas*'s narratives in CJ's body not only affects players' experiences of the gameworld's public space but also draws attention to this character's speech style, mannerisms, and built-in skills—his programmed habitus, which is "capable of generating practices regulated without express regulation" (Bourdieu 1977 [1972]:17). In my first play session, as CJ and I were trying to ride a bicycle, we accidentally stole a passing car. Pressing a single button on the controller yielded a complicated and physically difficult series of motions that CJ accomplished with ease and grace, throwing the driver onto the pavement and taking his place behind the wheel. Other characters often remind CJ of his abilities, saying "Even *you* must remember" how to do this or that. They draw on a memory the player lacks: "Takes you back, doesn't it, CJ?" someone asks after a high-speed BMX bike ride to escape enemy gang members. As I carry out my fieldwork in San Andreas, I can't interview CJ about his nostalgia for the drive-bys of his adolescence, but I can bring him into particular situations and experiment with his hidden capacities. CJ already knows how to fight, drive, dance, shoot, ride a bike, seduce a date, and talk trash. Notably, though—and in contrast to the previous GTA games—many of these skills improve with practice, implying that they are learned, rather than innate. At the start of the game, CJ is a terrible driver, gets short of breath when he exerts himself, and can't aim a weapon very well; as the player makes progress, the game controller becomes more responsive and CJ's physical capabilities gradually improve. Given that this engineering advance appeared in the first GTA game with a nonwhite lead character, it could represent a kind of safety measure: It limits the game's potential to reinforce racist conflations of blackness and natural-born criminality.

Encoded behaviors continually remind players that CJ has skills, physical capacities, memories, and inclinations different from their own. For example, the one-button carjack feature—common to all the GTA games—communicates the fact that the avatar is experienced with violence. This is a point of contrast with role-playing games, where players sometimes have to struggle not to "break character" by having their avatar act on their own knowledge. As Gary Fine writes, "The character does not have the right to know what the performer knows, and must maintain a pose of pretended ignorance about that primary self" (1983:241). But CJ need not pretend. He doesn't know who is at the controls, or even that there are controls. Meanwhile, the player can give CJ certain kinds of directives but cannot control the details of his story or the direction of his relationships with other characters; GTA's code presents a performance situation more like a scripted stage play than a framework for open-ended role-playing.

Digital game theorist Jesper Juul has offered a contrasting account of the one-button carjack: "In real life . . . entering a car is generally not considered a very interesting activity. . . . [S]ince entering a car is ultimately an uninteresting detail in the larger world of *Grand Theft Auto III*, the simulation of that activity is reduced to the pressing of one button" (2005:170–172). Here I would suggest that a reflexive, performance-oriented approach can make a substantial difference in game analysis. My own astonishment and pleasure at that first accidental carjack led me to think more about what it means for the avatar's competences to be radically different from the player's. It's safe to say that for the vast majority of GTA players, entering a car by throwing its driver into the street would indeed be a "very interesting" and unusual activity. The fact that it is a routine, automatic behavior for GTA avatars provides insight into the cultural world of the games, emphasizing that criminality is the norm. The practiced physical ease of the carjack also tells players that the identification between player and avatar is far from complete or transparent. As I provide the cues for virtual motions that I could not accomplish in my own body and did not design myself, I am aware of the perpetual tension between my performative agency and the encoded rules that govern CJ's every gesture—rules derived both from CJ's personal history and from the kinds of tasks that the player must accomplish in the course of any GTA game.

Deborah Kapchan suggests that the repetitive aesthetic practices that constitute "performance" tend to invite questions as to "what is reproduced, or imitated, and what is created and emergent." CJ's performative repertoire reproduces an actor's movements—through the medium of motion-capture technology—and continually displays what Kapchan calls "stylistically marked expressions of otherness" (1995:479). As in other digital games, this stylized physicality creates a palpable gap between the performance of the avatar and that of the player; following Kapchan, we might ask who is responsible for the on-screen performance, who is engaged in imitation or reproduction, and what is being imitated? Derek Burrill addresses this issue by describing an avatar's motions as "a type of choreography," which may reflect more general "cultural choreographic practices," particularly with respect to the representation of race and gender (2006:19; cf. Desmond 1993–1994). He notes that if the player leaves CJ standing somewhere without providing any input through the controller, CJ does not simply hold still: "A slight sway, breathing, and subtle head turning are accompanied by three ambient gestural sequences: shifting weight from one foot to another, bored and aching for some action; swinging of the arms from side to side in an effort to 'stay loose'; crossing and then uncrossing of the arms in a show of masculine bravado" (33). Left alone a bit longer, CJ sings snatches of songs heard on the San Andreas radio; the game camera zooms through the back of his head to represent a shift to first-person perspective and he begins to look around.

This eerie display of the avatar's independence from the player forges a connection between players and producers. The game designers are the ghost in the CJ machine, and their hauntings mark the player's experience as staged and prefigured. As Susan Foster writes about choreography, "One can see the residue of strategic choices concerning representation as distinct from the bringing to liveness of those choices. And in this distinctiveness, the contrasting functions of choreography and performance are apparent: dance making theorizes physicality, whereas dancing presents that theory of physicality" (1998:10). *San Andreas* often reminds players that someone else has choreographed this collection of physical markers of urban, black, male being-in-the-world; players collaborate with designers in bringing CJ's repertoire to life. In so doing, they participate in what W. T. Lhamon calls "the blackface lore cycle": "the choosing and the enlivening of gestures and style signatures" that perpetuates particular cultural codes (1998:69).

The theme of ethnographic complicity resurfaces here, in this case complicity with one's avatar. Like most interactive games, GTA offers "an invitation to comply or collude in the construction of a particular universe rather than in the deconstruction of its boundaries" (Burrill 2006:54). A player cannot explore much of San Andreas without actively supporting CJ's agenda. One cannot leave the city of Los Santos to investigate San Fierro until CJ has established his own gang territory by killing dozens of rival gang members. Similarly, one can only learn about gender relations in San Andreas by sending CJ on dates with various girlfriends, observing his protective impulses with respect to his sister's love life, or having him steal a lowrider and work as a pimp. This is an inverted, paradoxically one-sided form of complicity: The ethnographer must collaborate with CJ to be successful, but CJ doesn't know the ethnographer is there.

In effect, each player becomes the avatar's apprentice, building technical skills through increasingly difficult exercises until becoming accomplished enough to improvise. Again, we might surmise that GTA gameplay highlights the importance of participant-observation and of respecting local knowledge. The new player must learn how to enact what the avatar already knows how to do, be it driving through his neighborhood's streets, stealing new cars, playing darts, firing a weapon, or dancing at a club. As James Gee suggests (writing about digital games in general), "The player and the character each have knowledge that must be integrated together to play the game successfully," an instance of "distributed knowledge" in action (2006:177). The avatar has a programmed, unconscious repertoire of skills and behaviors, and the player must gradually acquire a parallel embodied knowledge of the commands required to animate him. In *San Andreas* there are also skills that the player and the avatar acquire *together*. With practice, CJ becomes a better driver and a better sniper. He gradually builds up increased strength and lung capacity through exercise, and he must take lessons to learn to fly a plane. These collaborative learning experiences can create a deep sense of identification with and loyalty to one's avatar—another aspect of ethnographic complicity, and one with particular ethical complexity when the avatar is engaged in violent criminal behavior.

THE MORALS OF THIS STORY: INTERPRETING *SAN ANDREAS*

In the fall of 2004, an MSNBC columnist greeted the release of *San Andreas* with this assessment:

> "Grand Theft Auto," the video game series reviled by parents for senseless violence and celebrated by gamers for pretty much the same reason, launched its latest installment last week. . . . Drive-by shootings, carjackings, crack use and

numerous other crimes which continue to make parts of America pure hell to live in have here been turned into fodder for entertainment. . . . No matter how good the game play, "Grand Theft Auto: San Andreas" is at its core a violent and mean-spirited game that rewards homicide, embraces racial and sexual stereotypes and uses real-life child-on-child violence as inspiration. (Loftus 2004)

A *New York Times* writer offered a similar account under the title "The Color of Mayhem":

The screen crackles with criminality as a gang of urban predators itch for a kill. The scene erupts into automatic-weapons fire in a drive-by nightmare of screaming car engines, senseless death and destruction set to a thumping rap soundtrack. . . . "They are nothing more than pixilated minstrel shows," said Joe Morgan, a telecommunications executive in Manhattan who is black and is helping rear his girlfriend's 7-year-old son, who plays video games. (Marriott 2004)

Such reports draw on the same rhetorical tropes as condemnations of gangsta rap. The writers focus on "senselessness," deleterious racial representation, and a threat to children (although all the GTA games are rated at least Mature/17+). These violent gangbanging narratives perpetuate stereotypes of urban black criminality, critics say, and they play into the hands of those who seek to profit from those stereotypes either ideologically or materially. As bell hooks puts it, "When young black males labor in the plantations of misogyny and sexism to produce gangsta rap, white supremacist capitalist patriarchy approves the violence and materially rewards them" (1994:122). Plantation labor and minstrel shows—these phrases are heavy artillery, and they suggest that those who make or consume these cultural products are participating in systemic racist exploitation. At best, then, we might view the stylized gangsta narratives of *San Andreas* as fabricated "fakelore" (Dorson 1976), a compilation of racist myths that misrepresent African Americans and are a damaging discredit to real urban lives.

There are others, though, who suggest that these critics have made the mistake of the Lion in the story of the Signifying Monkey: They have taken a story literally when it should be read figuratively. As Gates writes, "The import of the Monkey tales for the interpretation of literature is that the Monkey dethrones the Lion only because the Lion cannot read the nature of his discourse" (1988:85). A *Game Informer* review exemplifies this alternative reading of *San Andreas*: "A darkly bitter and unapologetic satire of modern society [that] brilliantly ridicules the hypocrisies of our culture through heavy-handed shock value and subtle language choice. . . . The outrageous criminal actions you commit will undoubtedly spark the controversy: 'How can they make a game that lets you do this?' Meanwhile, the game itself asks a more complicated question of its players: 'How did we end up with a society that allows these things to be satirized in the first place?'" (Miller n.d.). Critics who fail to acknowledge the game's ironic character risk being branded "yokels," like the nineteenth-century audience members who occasionally misread minstrel-show blackface as genuine blackness: They "never saw the doubleness at all" (Lhamon 1998:173).

The Rockstar staff borrowed *San Andreas*'s Signifyin(g) aesthetic—as well as many voice actors, creative consultants, and musical tracks—from hip-hop, from independent "ghetto" films of the 1980s and '90s, and by extension from earlier blaxploitation films and black music genres (Demers 2003). Like many rappers, *San Andreas*'s characters often channel the black vernacular archetypes of the badman, the trickster, the Fool, and the Devil, all expert Signifiers (Quinn 2005:92–140, Wheeler 1991:213). The aesthetic, ideological, and commercial success of gangsta rap and of games like *San Andreas* all depend on this double-voiced quality, which allows producers, listeners, and players to walk the line between claiming these cultural artifacts are unimpeachably authentic—because they represent bleak realities of violence, poverty, and lack of opportunity—and that they are allegorical, ironic, not to be taken literally: just a game, just a song, just a boast, just a tall tale, a "children's story" made for adults.

The transgressive violence and sexuality in *San Andreas* is accompanied by constant self-aware references to out-of-control consumer culture and to gangsta rap itself. On the gangsta rap radio station "Radio Los Santos," a commercial for "Ice Diamonds" blares at CJ: "Nothing says 'I love you' like a lump of carbon mined by wage slaves in Angola. . . . Passion! It *can* be purchased. And it can be overpriced." Or consider CJ's friend Jeffrey, a teenaged aspiring rapper of dubious skills. Bedecked with gold chains and tattoos, slinging street talk with self-conscious anxiety, he makes the real Grove Street gangbangers shake their heads in pained amusement—and becomes a commercial success across San Andreas. This character is a self-made minstrel-show rapper of the type bell hooks critiques, but he is presented as pathetic and ridiculous; his buffoonish existence serves to emphasize the fact that CJ and his friends are *not* caricatures of this kind (see Chapter Two).

San Andreas also Signifies upon the culture of violence it is often accused of celebrating. In one cut-scene, CJ is talking to his Chicano friend Cesar about the need to keep a low profile on an upcoming mission.

CJ: I was thinkin' about welding me some shit together and making a silencer.
CESAR: You're fucking crazy, holmes. You've gotta get out of this ghetto mentality.
CJ: So what *you* thinkin'?
CESAR: Let me show you. Check *this* out, holmes. [*Gets out silencer.*] Here, take mine.
CJ: Where'd you get that?
CESAR: Same place I bought my pants, holmes. This is America.
[On my field recording, the person letting me observe his gameplay erupts in laughter.][12]

This exchange gains new resonance when read against comments by Lazlow, a GTA writer and professional technology critic: "You can go to Wal-Mart and buy a shotgun, but they force record companies to bleep out profanity in CDs? If you're in a dark alley, who are you afraid of, the kid with a video game or the one with the shotgun?" (Patterson and Lazlow 2002).

Like folktales and fables, the GTA games engage players' attention through intertextual references that rely on prior experience and established values. While *San Andreas* is the first GTA game to be set in an explicitly hip-hop-derived cultural milieu, all the games have featured hip-hop on their radio stations. As with my opening example from Slick Rick, these songs present parallels to the games' narrative content and their stylized representation of violence. Players and critics who are familiar with hip-hop culture often recognize GTA's ironic spirit and its pervasive hip-hop citations. Black comedian Dave Chappelle—much of whose work revolves around race issues and the hip-hop world—won the hearts of many GTA players with his satirical live-action reenactment of a *GTA III* carjacking; he highlighted the thoroughly artificial and over-the-top nature of the game's violence, undermining the argument that GTA is dangerously realistic (Chappelle 2003). Three years later, after the release of *San Andreas*, Chappelle created a sketch in which DJ ?uestlove plays a newly discovered song by deceased West Coast rap icon Tupac Shakur—a song ostensibly from 1994 but that contains prophetic references to George W. Bush's election and the bombing of Afghanistan. Chappelle's Tupac takes care to name-check Carl Johnson: "We'll make you bustas pay us, /Run up in your spot like CJ from San Andreas" (Chappelle 2006). When CJ's doings are narrated by a famous hip-hop MC from beyond the grave, the game's intertextuality comes full circle.

A *Washington Post* reporter recounts another reception story, based on interviews with teenaged Chicano players from South Central L.A. (who were playing *San Andreas* in a subsidized-housing bedroom decorated with Tupac Shakur and *Scarface* posters).

> "It's a *game*, just a game, right? But at the same time, it's *more* than that. There's reality to it," says Tito. . . . "Even down to the choppy Spanglish, the 'Ora le, homes,' that some of the gangstas say," Tito goes on, "it's all realistic." The other guys who aren't in South Central "won't fully understand. For them, it's just entertainment." . . . "San Andreas" is like a fun house mirror to Tito, an exaggerated yet still realistic version—the dueling gangs, the racial tension—of his everyday life. (Vargas 2005)

While these young men acknowledged the stereotypical qualities of the game—"The game's violent, yeah. It's dangerous, yeah. It's a stereotype, yeah," said one avid player—they also appreciated its primarily black and Latino characters and its reproduction of their own neighborhood, music, and speech styles. As in hip-hop reception, where some interpreters see offensive racial caricatures, others see stock characters from an established expressive tradition. Both gangsta rap and *San Andreas* employ the "exaggerated features" typical of folktale characters, features that "set them apart from reality" and "enhance with clarity the necessity and probability of any given situation" (Kamenetsky 1992:82; cf. Quinn 2005).

Game scholar Kurt Squire offers another perspective, gleaned from interviewing low-income African American high school students about their experience with *San Andreas*. Squire reports that "these kids found it somewhat bizarre that we would ask them about violent video games . . . when there were clearly many more tangible causes of and forms of violence in their lives (i.e., poverty, drugs, lack of economic mobility)." He asked the students what they thought was unrealistic about the game; they responded that "the way that blacks could save to buy a house and eventually move to the suburbs was the 'most unrealistic.' . . . These players seemed able to construct a fairly serious critique of the current socioeconomic order in the United States, developing *through play* seedlings of a structural theory of how race and class are reproduced in contemporary America" (Squire 2008:177).

Of course, not all players are equally attuned to GTA's ironic dimensions or equally familiar with the realities of structural racism that *San Andreas* depicts. Moreover, GTA's parody is backed up with gameplay engineering designed to create visceral excitement based on violence, speed, and mayhem. As with gangsta rap, the reflexive cultural critique is there for those who are interested, but it is difficult to argue that this is what makes these games sell millions of copies. But I want to contend that this double-voiced quality itself is part of what makes GTA so compelling to players. Its ambiguous implications mirror the themes of freedom and constraint explored in the gameworld. In theory players can do whatever they want in San Andreas, but in practice CJ constantly experiences pressure from different quarters. The player's tasks typically involve using limited resources in creative ways in order to respond to those pressures—the "ghetto mentality" of welding scrap materials into something useful. Players engage in a similarly satisfying creative endeavor when they work at identifying and interpreting the scraps of political and pop-culture references that have been cobbled together to build the state of San Andreas. Ambiguity is a fundamental attribute of play, as Sutton-Smith observes; in his terms, *San Andreas* straddles the continuum between "play" and "metaplay," combining hardcore gaming engineering and constant comedic metacommentary (1997:147–150). Players acknowledge and often celebrate this ambiguity in order to make sense of *San Andreas*.

Both GTA's players and its makers have a great deal invested in creating a sense of community around the series, encouraging "collaboration in a distributed medium" (Kirshenblatt-Gimblett 1996:30). People may play GTA in private, but they do not play alone; they remain aware of the existence of millions of other players. The game's interpretive communities are geographically dispersed, but they have all spent time in one virtual world (the state of San Andreas) and they meet, debate, and share material culture in another (the variegated public sphere of the World Wide Web). Rockstar Games encourages their activities by supplying official screenshots and advance information to gamer websites, building speculative buzz for new games. The company also listed 12 player-produced GTA websites at the back of the *San Andreas* instruction booklet—including a site specific to *San Andreas*, which was evidently thriving well before the booklet was printed and the game was released. Incentives for visiting player-produced sites include gaining special knowledge (user guides, maps, cheat codes, "easter eggs"); sharing one's own virtuosity in mod design, avatar construction, or gameplay; and reading and writing tales of exploits and frustrations in the gameworld (often supported by screenshots). As Matt Hills has shown, such forums rely on the assumption that participants are not just consuming the same product but can experience "a common affective tie" (2002:180). However, while many GTA fans do seek each other out online to confirm that tie, others are disgusted by the language used in certain forums or feel it is a point of pride to solve the game's riddles on their own. Because GTA is not a multiplayer online game, players can easily separate themselves from those who play "for the wrong reasons."

Negative media attention plays a crucial role in constructing interpretive communities of players, because it tends to draw fans into a defensive collective. But player responses to media critiques also reveal deep ideological rifts. For example, players express widely divergent views about racial representation in San Andreas and how it has been assessed in mainstream publications like the *New York Times*. They hash out these differences of opinion in heated debates on gamer forums and in comment threads for newspaper articles. Such debates began from the moment that *San Andreas* preview material revealed that the lead character would be African American. Whereas newspaper reviewers cried foul over gangsta stereotypes, game industry reviewers and players noted that Rockstar also risked alienating white players who had never before been compelled to use a nonwhite avatar.[13] Their predictions were borne out by gamer forums in which many participants complained that they did not want to play as a black man—prompting antiracist rebukes from others (e.g., gtagaming.com 2004).

After the game's release, players responded to the *New York Times* "Color of Mayhem" article (Marriott 2004) in many online venues. On PS3Forums.com, for example, a forum administrator suggested, "If anything the makers of San Andreas are providing us with Racial equality. The first few games you played as a white person. Now as soon as you play as a black person people get all huffy about it." Another contributor—"Dragonlance," who identified himself as a Russian Jew living in Arizona—criticized perceived black privilege (e.g., in university admissions) and was warned by the forum administrator that his comments were offensive. This exchange prompted a response from a "black canadian and a really big fan of GTA," who made it clear that s/he did not appreciate Dragonlance's remarks but also did not assess *San Andreas* as racist. This individual's take on racial representation in the game relied on knowledge of its sampled sources:

> The game is not trying to discriminate or stereotype black people, it's just trying to show the black point of view of urban style gangster life. The game is just taking the gangster movies like Boyz In the Hood, and Belly, and putting them in playable format, so i dont know why there is such an outcry from the black american community. but then again i am canadian and up here i guess there is alot less racism than in the usa, if you ever come to toronto you will be seeing alot of interacial couples walking around. (PS3Forums.com 2004)

Similar debates played out in 2007, after a *Black Voice News* writer criticized *San Andreas* for its potential bad influence on minority youth (Jones 2007). Again, some participants responded with racist vitriol, others mounted complex defenses of the game's intentions, and a few identified themselves as black and asserted that the game employs stereotypes but is not racist. Some of these latter postings met with angry responses, for example, "If you're black, and you're sitting here saying GTA SA wasn't exploitation, then you're clearly an oreo cookie uncle tom. Either that, or you ARE one of the idiots who think thug life is meaningful" (PS3.QJ.net 2007).

Such discussions demonstrate that there is no monolithic "black perspective" on *San Andreas*—nor is there a unified white, Latino, queer, female, or parental perspective. When players do opt to identify themselves as members of one of these groups, their diversity of opinion points to the importance of interpretive alliances that transcend these basic demographic categories. GTA is always played and interpreted "in a situation," a situation with a particular institutional context (Fish 1980:309)—in this case that of mass-produced popular culture and mass-media critical reception of those products, against which certain audiences can position themselves as marginalized, misunderstood, or intellectually superior. Game designers and critics sometimes idealize total immersion as the ultimate goal of gameplay, hearkening back to Huizinga's notion of a "magic circle" dividing game experience from everyday life (Huizinga 1955; cf. Juul 2005:164–165). But as Martin Barker and Kate Brooks have shown, "practices of pleasure" connected with media reception rely on "a pattern of involvement which extends beyond the moments of pleasure" (1998:145). Digital-game ethnographers should treat these "patterns of involvement" as an essential part of the gameplay experience, be they private thoughts, collective discussions, or creative practices like the development of software mods and the distribution of GTA-related videos on YouTube. By doing this interpretive and inventive work, players are drawing up a portion of the game rules; they are developing their own convictions about what counts as satisfying, successful, and ethical play.

I have belabored the importance of ambiguity and interpretive agency in this chapter because so many critical appraisals still treat video games and other popular media forms as direct instruction or sinister brainwashing—particularly when their subject matter revolves around violence, sexuality, and gendered/racialized characters. Video games have come in for special criticism because of the possibility of long hours of immersion in the gameworld, the repetitive nature of many game tasks, and the player's active performance of violent behavior (as opposed to a moviegoer's spectatorship, which is assumed to be relatively passive and detached). Consider the comments of Congressman Joseph Pitts (R-PA) at a June 14, 2006, hearing of the House Subcommittee on Commerce, Trade and Consumer Protection: "It's safe to say that a wealthy kid from the suburbs can play *Grand Theft Auto* or similar games without turning to a life of crime, but a poor kid who lives in a neighborhood where people really do steal cars or deal drugs or shoot cops might not be so fortunate."[14] Statements like this one lay bare the political stakes involved in research on digital games.

Without a doubt, digital games do affect the values, beliefs, and cultural common ground of their consumers. But they are not like the programming code that governs the behavior of inherently compliant video game avatars—or, in Pitts's view, poor inner-city kids (cf. Gauntlett 2001:57). Rather, digital gameplay is a form of expressive culture developed through collaborative performance and intertextual interpretation. Like storytellers or actors, players move CJ through his episodic travails. He is not their creation, and they don't necessarily identify with him—any more than storytellers always identify with their protagonists or ethnographers with their subjects. Different players bring different "portfolios of interpretation" to the material, building up their own forms of vernacular culture and normative practices around these artifacts (Hill 1997). Media accounts that deny this diversity by insisting on the totalizing "bad influence" of the GTA games inspire players to trade interpretations and compare practices, validating the freedom of their own play while also confirming points of commonality.

In the 1970s, Michel de Certeau asserted that consumers of mass media products are not simply "grazing on the ration of simulacra the system distributes to each individual," but he also mourned that the expansion of systems of production "no longer leaves 'consumers' any *place* in which they can indicate what they *make* or *do* with the products of these systems" (1984 [1974]:xii). Ordinary people are engaged in creative interpretation, elaboration, and criticism of myriad texts, he claimed, but have no venue for displaying evidence of their work. Things are different today. The wealth of material that has grown up around *Grand Theft Auto*—the discussion forums, screenshots, mission guides, maps, mods, machinima, and cut-scene transcriptions that circulate around the world—demonstrates that digital gameplay is a globally dispersed form of participatory culture that is generating new aesthetic and interpretive traditions (Jenkins 2006b). These games also tend to collect and transmit preexisting cultural narratives and modes of expression drawn from both contemporary popular media and more traditional folkloric domains.[15]

As we consider the emerging performance practices of digital gameplay, thinking through the family resemblances among gameplay, tourism, and fieldwork might enrich our understanding of players' experiences and interpretive work. In this chapter I have argued that some of the analytical habits of mind and the satisfactions of playing these games closely resemble those associated with ethnographic fieldwork—not only old-paradigm imperialist ethnography but also newer models, in which the ethnographer aims to develop collaborative relationships with research associates (the avatar, nonplayable characters, other players, and the game designers). For me, doing GTA fieldwork reinforced the lesson that neither ethnographic field sites nor single-player digital gameworlds are closed-off "culture gardens" bounded by "magic circles," though thinking about them that way yields some interpretive insights. Just like ethnographers, video game players bring conscious theories and strategies as well as subconscious cultural knowledge to their gameplay; no one does fieldwork or gamework in a closed system. Because Rockstar's game design always blends immersion-enhancing realism with immersion-disrupting parody and citation, these games keep each player in the liminal state of "playing along" that partially defines the classic fieldwork experience: not a tourist, but not a local; trying to act naturally while consciously storing away new knowledge; in the world but not of the world.

CHAPTER 2

Jacking the Dial

Radio, Race, and Place in San Andreas

In December of 2005 I watched and took notes as a *Grand Theft Auto* player carjacked a convertible that looked like it might be suitable for a pimping mission. The clean-cut white guy behind the wheel was in the wrong part of town. CJ, the player's avatar, tossed the driver from the car and jumped in. As he tried out the lowrider's hydraulic controls, the DJ on the local country station spoke through the car stereo: "It's like Kant said: You can be an active originator of experience or a passive recipient of perception. I tell the fellas behind the diner that all the time" (gameplay notes, December 17, 2005; Kant 1998 [1781]). When the next country classic started, the player flipped to the modern hip-hop station and stepped on the gas.

Five years later, after YouTube had become an important site for music distribution, the following item popped up in my Twitter feed:

> Top comment on YouTube vid for Ralph Tresvant's Sensitivity: "man I killed so many people to this song on Grand theft auto San Andreas" (posted August 15, 2010 by islandis, retweeted by wayneandwax)

When I searched for this song on YouTube, I found comment threads full of nostalgia—some for the actual 1990s and some for 1992 San Andreas (e.g., soulfulbohemian 2008).

While the *Grand Theft Auto* series has attracted considerable media attention for its violence and sexual content, few have explored the role of music in its appeal—despite the fact that players and industry reviewers often applaud GTA's musical soundtrack system as an innovative design feature that improves gameplay immersion (Spence D. 2004). Most digital games have nondiegetic soundtrack music, which functions like a dynamic, adaptive film score: Audio programming shifts seamlessly among precomposed segments of music that suit particular gameworld places, activities, narrative episodes, or emotional states (Whalen 2004, Collins 2008). The GTA games are different. All of the gameworld music is diegetic, played by in-game radio stations that can be heard only in places where the games' characters might plausibly find radios: in the gameworld's retail establishments, in the homes and cars of the avatar's allies, and in the cars that the player steals. On foot, the city soundscape consists of only traffic noise, the comments of passersby, and occasionally the radio stations selected by nonplayable characters. When players are driving, they can choose the music to accompany their exploits—a significant incentive to steal a vehicle and take long drives around the gameworld. By withholding conventional soundtrack music and giving players control over the radio system, the game designers draw attention to music's capacity to channel emotional experience, match or clash with narrative situations, and alter the passage of time.

In *Grand Theft Auto: San Andreas*, the radio stations play licensed tracks spanning genres from country to house to hip-hop, all chosen to suit the game's 1992 West Coast setting (▶ 2.1). The songs on the radio are introduced by DJs and interspersed with parodic advertisements; for example, a home-security ad for the "Executive Intruder Extermination Service" features an anxious upper-class mom consoled by an announcer who proclaims, "We'll ensure that you'll live in a fortress!" GTA also parodies many real brands, including Burger King, De Beers, and Nike. But the only direct product placement in the games is the music on the radio, a design choice that creates a powerful musical connection between the gameworlds and the real world. Players often comment on the "realism" of this soundtrack system; as one man wrote in a survey response, "It fits in with the more gritty feel of GTA compared to other games—the stuff going on isn't something that merits an orchestral soundtrack, it's just stuff that happens while driving around" (David Newgreen, Melbourne, Australia, 18–21).

This chapter investigates the musical choices made by GTA's designers and players. Era-appropriate musical tracks and the radio format in which they appear play a key role in all the *Grand Theft Auto* games. The player-controlled radio stations not only increase the verisimilitude and immersive qualities of each gameworld, but also encourage players to associate particular music with particular characters and places. Moreover, in creating radio playlists that purport to represent classic country, reggae, rock, house, funk, hip-hop, and other genres with 15 or so songs each, the game designers are constructing canons. Their playlists have instructional force, teaching millions of players how to recognize and value certain artists, songs, and musical characteristics within a given genre. *San Andreas* also dramatizes various kinds of encounters with the commercial music industry, including the perspectives of the casual radio listener, the gangsta rap connoisseur, and the aspiring hip-hop artist. The game's protagonist is not a musician, but music shapes his personality and his terrain. Gangsta rap production is presented as both an avenue of upward mobility for young black men and as a cynical commercial enterprise that may entail sacrificing one's street cred and artistic integrity.

In general, GTA gameplay creates an engaging tension between the pressure to conform to the standards of the avatar's social milieu and the appealing freedom of driving around a complex landscape, spinning the radio dial, and engaging in atypical social behavior (ranging from violence, larceny, and pimping to getting fat on fast food, running red lights, and dressing in costume). In every version of the game, the radio invites players to explore the role of music in their navigation of an unfamiliar landscape and their occupation of a foreign persona. Rather than giving the player cues to follow (e.g., ominous music when danger is approaching), the games allow players to use music to channel their own emotional states. As I discussed in Chapter One, GTA's "freedom" derives partly from its mediated, artificial qualities: Players can control their degree of identification with local characters, choose their own preferred depth of immersion, and tack between viewing the games as consequence-free escape zones or as trenchant commentary on other violence-saturated media products. The in-game radio stations facilitate this process. Players use the real-world music, gameworld-oriented DJ patter, and parodic advertisements to shape both their sense of place in the gameworld and their ethical identification with the criminal characters they occupy.

Grand Theft Auto players are a diverse and far-flung group—there are many fan collectives, but there is no unified GTA subculture (see Chapter One). But this fact need not completely overwrite players' common experiences, based on their separate but parallel performances of particular stories in particular places. In terms of duration, agency, narrative constraints, character identity, and playfulness, GTA creates a different kind of common experience than that shared by the dispersed audience for a Hollywood release or by all the tourists who have visited the Eiffel Tower. The games' virtual spaces also frame musical reception in relatively specific and constraining terms. In the actual world, the latest hit single might reach listeners as an overplayed radio track, club music for dancing, a legal or illegal download, a CD, the audio component of a music video, and so on. But the music of *GTA: San Andreas* reaches players as they actively navigate a particular digital city while occupying the virtual body of a particular 20-something black man from "Ganton" (based on L.A.'s Compton).

In 2005 I created an online survey that asked players about their experience with the GTA radio system (Miller 2005). Survey respondents were self-selecting; I initially recruited respondents by posting the survey link on gamer discussion boards. New waves of responses appeared in my inbox when the survey was discussed in several blogs (e.g., Thompson 2006) and when it was mentioned in the comment thread for a FARK.com posting on video game music (FARK.com 2006). I received 92 survey responses, 88 of them from men. Occupations ranged from librarian to mechanic to film editor, but 55 percent were students and 18 percent worked in information technology (including digital-game design). They hailed from at least seven countries— the United States, Canada, the United Kingdom, Australia, Germany, the Netherlands, and Poland—but the majority were from North America. Their ages were fairly evenly distributed from the late teens to the late 30s. Virtually all respondents owned more than one GTA game, and the vast majority owned and had played all the way through *Grand Theft Auto III*, *Grand Theft Auto: Vice City*, and *Grand Theft Auto: San Andreas*. Completing the plotline of each game requires about 40 to 100 hours of play, depending on the game edition, the player's skill, and his or her interest in optional side missions and unstructured exploration. My survey respondents can be considered seasoned travelers in the GTA gameworlds, with a substantial common knowledge base. This chapter draws on their accounts of their experience with the GTA radio stations as well as on the broader range of sources already cited in Chapter One (online discussion boards, published game reviews, gameplay observation sessions, interviews with players, and my own gameplay experience).

GRAND THEFT AUTO'S GAMEWORLDS AND RADIO STATIONS

Six major *Grand Theft Auto* games were published between 1998 and 2008, along with an assortment of expansion packs and editions for portable game systems. Three of the main releases stand apart as a loosely related trilogy: *Grand Theft Auto III* (2001), *Grand Theft Auto: Vice City* (2002), and *Grand Theft Auto: San Andreas* (2004). The original *Grand Theft Auto* (1998) and *GTA II* (1999) set up the general game concept, but *GTA III* made the radical shift from top-down two-dimensional graphics to a 3-D urban landscape that could be viewed from a variety of camera angles controlled by the player (cf. Strain 2003 on perspectival shifts across the history of cinematic film). *GTA III* vastly outstripped its predecessors in terms of sales, and it established visual and narrative conventions that were retained and developed in *Vice City* and *San Andreas*. See Table 2.1 for a chart of these three gamespaces and their radio stations.

The production personnel employed by Rockstar Games varied somewhat across these three games, but the most senior figures always included the Scottish brothers Sam and Dan Houser and their British school friend Terry Donovan (three of Rockstar's founders, all of whom had previously worked for the media conglomerate BMG). Craig Conner and Stuart Ross wrote and produced original music for the radio stations (e.g., genre-parody pop songs and

Table 2.1 *GRAND THEFT AUTO* RADIO STATIONS

game	gamespace	radio stations	genres
GTA III (2001)	Liberty City (New York, 2001)	Head Radio	"nonstop rock"
		Double Cleff FM	opera
		K-Jah	dub
		Rise FM	techno
		Lips Radio	"disposable"
		Game FM	hip-hop
		MSX FM	drum'n'bass
		Flashback FM	*Scarface* ('80s pop)
		Chatterbox FM	talk
Vice City (2002)	Vice City (Miami, 1986)	WildStyle	early hip-hop
		Flash	top 40
		K-Chat	talk
		Fever 105	R&B
		VRock	hard rock/metal
		VCPR	faux NPR
		Radio Espantoso	Latin
		Emotion 98.3	soft rock
		Wave 103	new wave
San Andreas (2004)	San Andreas (California/ Nevada, 1992)	Playback FM	"classic hip-hop"
		Radio Los Santos	"modern hip-hop"
		Bounce FM	funk
		K-Rose	country
		K-DST	classic rock
		SFUR	house/techno
		K-Jah West	"dub/reggae"
		CSR 103.2	contemporary soul
		WCTR	talk
		Radio X	alternative rock
		Master Sounds 98.3	classic soul

Genres in quotes are taken from game liner notes or from on-air station IDs. Stations are listed in the order in which they appear in the liner notes.

commercial jingles), while the American radio personality and technology critic Lazlow did much of the scriptwriting for DJ banter and commercials. Lazlow and Dan Houser were responsible for the overall design of the radio system. Game-development companies work under grueling schedules to bring games to market as rapidly as possible; under the circumstances, Rockstar's investment in the GTA radio suggests it was considered a high-value component of the games. The company's most senior staff were closely involved with radio production work, to which they brought their previous professional experience with music video direction and record label management (Design Museum 2002). Many employee hours were spent selecting and licensing era-appropriate music, writing DJ patter and commercials, and organizing hundreds of voice actors.

The GTA games are full of dark humor, intertextual citations, and political parody (see Chapter One). In-game commercial radio and billboards are the primary channels for the games' critical commentary; as such, they constitute GTA's political voice. As an Australian player observed, apparently recounting a GTA ad script off the top of his head,

> The ads and commercials seem very sarcastically patriotic, for example,
> "Mike, I can't feed the kids, and the rent's due!"
> "Are you saying this isn't the greatest country in the world?"
> "But Mike, my . . ."
> "Hold on a second. Everybody, USA! USA! USA!"
> (Justin Grima, Victoria, Australia, 18–21)

This ad takes place at a book-tour event for a conservative self-help author, apparently in a football stadium or similar setting. The book, titled *Rags Are Riches,* purports to teach the poor how to appreciate their poverty. In the ad, the author tells a homeless man, "Instead of complaining about being poor, enjoy it. Watch TV. Don't vote. Who cares?" A *game-brains.com* reviewer cited this ad to support his argument that *San Andreas* is "among the most politically engaged pieces of mainstream art to come along in the last ten years" (LaVigne 2005).

The GTA gameworlds, their radio stations, and their avatars' personalities all grew more complex across the three games under consideration—as one would expect, since Rockstar needed to maintain player interest in the series. As company president Sam Houser said of the design process for *San Andreas,* "Our goal was to develop the key themes of Grand Theft Auto: freedom of choice and the ability to do lots of things closer to their logical conclusion; a bigger world, with much more to do in it; and much more interactivity between you and the environment, between player and lead character, and between characters within the game world" (2004). A brief survey of the games will serve to illustrate this design trajectory, which was characterized by ever-stronger connections between the avatar's character, his local terrain, his era, and what's on the radio.

GTA III is set in Liberty City, a thinly veiled New York City. The game's time period is roughly contemporary with its 2001 release; Liberty City lacks the deliberately nostalgic period markers that are scattered throughout Vice City and San Andreas (such as comically gigantic cell phones). Liberty City's nine radio stations also contain far less place- and era-specific music than in subsequent games. Many of the songs were written and produced specifically for the game; they are genre parodies of sound-alike commercial pop. The dub, techno, and drum 'n' bass stations do feature a small number of real-life artists, but not in a realistic format—for example, all the tracks on K-Jah are licensed from a single 1981 album by the dub artist Scientist.[1] Flashback FM adds a little nostalgic depth to the lineup, as its name suggests, but its playlist consists exclusively of tracks from the 1984 *Scarface* soundtrack. Moreover, as one player pointed out to me, the inclusion of drum 'n' bass, techno, and dub stations tends to mark the whole radio dial in this game as more British than American. In general, this material is not particularly representative of what was on the radio in New York at the turn of the millennium.

Despite its uneven verisimilitude and shortage of licensed tracks, the *GTA III* radio does establish some key features that persisted in subsequent editions. The radio provides comic relief and alleviates what Frasca has called "errand boy syndrome," the tedium of repeatedly traversing the same gamespace in the course of completing different missions (2003). Lazlow's surreal political satire on "Chatterbox," Liberty City's talk-radio station, attracted a huge following and set the standard for the political content of subsequent games. Still, the radio in *GTA III* does not engender the intensely local and intimate sensibility that makes real radio such a powerful medium (Berland 1994). The glib irony of both this gameworld and its radio dial undercut the game's capacity to create an immersive and dramatically engaging experience. Even the avatar is an empty shell—a generic small-time crook of unspecified white ethnicity, he has no name or voice.

In *Vice City*, the player's avatar has a bit more depth: he is Tommy Vercetti, a young mobster recently released from his first major stint in prison and deployed to Vice City by his Liberty City bosses to expand the family business. Tommy is voiced by actor Ray Liotta, and he not only converses with other characters in "cut-scenes"—filmlike segments that advance the game's narrative—but occasionally expresses internal thoughts in voiceover. Tommy has an ethnicity, if not much of a personality. As for the *Vice City* radio, it was produced by much of the same personnel as *GTA III,* but the end product is remarkably different. Vice City is 1986 Miami, and its music is place- and time-specific from the start. The man who picks up Tommy Vercetti at the Vice City airport is listening to a salsa station. The first time Tommy drives a car, the radio is playing Michael Jackson's "Billie Jean." Vice City's nine radio stations play almost 100 era-appropriate licensed tracks, from Twisted Sister to Grandmaster Flash to A Flock of Seagulls, all organized into fairly realistic commercial-format radio playlists. A Spanish-language station, Radio Espantoso, represents a major step in linking the radio to the game's locale. There are only five original songs on the radio: three genre-parody Latin tracks and two songs credited to the fictional Scottish hair-metal band "Lovefist" (a kissing cousin to "Spinal Tap"). Lovefist's members are characters in the game; their songs appear amid the licensed 1980s hits on the game's hard rock station, creating a musical bridge between Vice City and the actual world. (Lovefist also makes an appearance late in the storyline of *San Andreas*, linking the two gameworlds.)

The *Vice City* radio was a much-beloved aspect of the game, spawning a seven-CD soundtrack box set of 1980s hits. It evoked nostalgia in at least two generations of players: those who were teenagers in the mid-1980s and a younger cohort who remember many of these tracks as the music on their parents' car radios. (Today, these songs also evoke nostalgia for playing *Vice City*, as the example at the beginning of this chapter illustrates.) A 20-something player explained the attraction of *Vice City*'s satirical 1980s gameworld in an interview:

PLAYER: My mom is still in the '80s. She still wears the leather jackets with the bad hair . . . I don't know, I have this idea of the '80s as like, you know, the heavy metal bands, and they wore the really tight, the skin-tight pants that are all pinned up the sides, with the tight shirts with the big collars—that's the '80s, to me.

[We discuss "Lovefist," who fit this description perfectly.]

PLAYER: I love the '80s! They're funny! I was born in the '80s, I'm supposed to love them.

KM: So, there's a big kind of nostalgia factor.

PLAYER: Yeah, I guess so—I don't remember the '80s at all, but, I don't know, I find I'm always attracted to, I like things that are cheesy, or that would normally be considered cheesy now.[2]

This player's delight in *Vice City*'s campy, flashy aesthetic was typical of my survey respondents, many of whom noted that they were too young to actually remember 1986. As one man wrote,

> Hearing the '80s music [in *Vice City*] recalls for me not the '80s specifically, but representations of the '80s—music videos, movies, and tv shows. These are all images that have been mediated and feel artificial to me, so the music helps to enhance the feeling of, not artifice, but playfulness in the game. The music emphasizes that what I am experiencing is not serious, so I can enjoy myself and do whatever I feel like, to my character and to the other characters of the game. (Jesse, Nova Scotia, Canada, 22–26)

In a game whose plot is largely oriented around contract killings, a 1980s pop soundtrack and ads that skewer Reagan-era conspicuous consumption provide a welcome counterweight to stylized brutal violence.

But in terms of creating an immersive environment and a distinctive gameworld, *Vice City* is almost too successful in its mimicry of commercial radio. The stations serve up such a broad range of 1980s hits that their nostalgia appeal becomes generic. Largely disconnected from Tommy Vercetti's character and circumstances, the radio speaks directly to the player on the other side of the screen. Many survey respondents expressed their affection for the *Vice City* radio and praised its ability to capture an era, but no one reported any relationship between the radio and their identification with the avatar.[3] That identification was not particularly character-driven in any case; as one player explained in an interview, "[Tommy's] personality actually doesn't come across . . . He doesn't have a whole lot of personal style. His clothes are all purchased by other people, his cars are all stolen, he stays in a hotel, so you don't see any of his personal items."[4]

In *GTA: San Andreas,* by contrast, the player's avatar comes with family, friends, and a substantial backstory (see Chapter One). To recap: Carl Johnson (CJ) is a poor, young black man in 1992 Los Santos/Los Angeles. Corrupt police officers have framed him as a cop killer so that they can manipulate him as they wish. Meanwhile, on the home front, CJ's older brother often goads him with accusations that he abandoned the Grove Street Families gang during a five-year stint on the East Coast. CJ takes up residence in his childhood home, where we see him grieving over a picture of his murdered mother. His friends and brother still live in the neighborhood where he grew up. When he visits their homes, they give him missions, many of which involve reclaiming Grove Street Families territory from drug dealers. All this makes CJ a very different kind of avatar from Tommy Vercetti.

CJ actually seems too real to some players; the race and class issues raised by the game hit too close to home to be funny. In an interview, a Canadian woman in her 20s explained why she laughs at the gender stereotypes in *Vice City* but is less comfortable with the gangsta scenario in *San Andreas.*

PLAYER: [In *Vice City*] I laugh at it because it seems really self-conscious. Like all the women that are always wearing bikinis. It's sort of like, this is fun and we're going to put it in, because mostly guys are gonna play this game, but we're going to take it a little far so that it's not quite serious. . . .

KM: But then the thug aspects of San Andreas don't seem to strike you the same way.

PLAYER: Yeah, well, I don't know, it might just be that I don't like the game as much. But, yeah, it seems a little, a lot more serious, especially with how popular R&B and things like that that are dominated by an *idea* of black culture. This sort of seems to be trying to actually fit into that, and actually *be* that and *represent* that sort of idea. It's like watching music videos of the guys with their big black Expeditions and their gold chains. Yeah, it seems like it's trying to be more realistic. . . . Whereas I mean Vice City is a lot easier to be a parody, because it's of the '80s, and the '80s were funny. [*laughs*] . . .

KM: So, do you feel like you identify with the avatar when you play? With his backstory and character?

PLAYER: Tommy? No, not really.

KM: And CJ, to the extent that you've played with him?

PLAYER: No, not really CJ either, except for that I find—if anything, I have less of an association with his community and his motives and things. But then there are other things that he cares about, like he actually cares about things, whereas Tommy is sort of more of a straight-up thug, and he just wants the money back for revenge and honor and pride of place. Whereas CJ is sort of more motivated by, oddly enough, the emotional aspects of his associations with people.[5]

▶ 2.2

Like this new avatar, the *San Andreas* radio dial has a more defined, more complex, and in some ways more constraining character than in previous GTA games. Of the 11 radio stations in CJ's world, 7 play hip-hop, soul, funk, dub/reggae, and house—genres intimately connected with urban black culture.[6] This music facilitates identification with CJ and his history, drawing the player into the gamespace instead of broadcasting out into the living room. The hip-hop stations have particular resonance, since the four classic elements of hip-hop culture make appearances in *San Andreas*: DJs, MCs, graffiti, and competitive dance all play important roles, along with the sometimes-mentioned fifth element of sartorial style. Graffiti appears all over the gameworld; CJ can spray his own tags over those of rival gangs. He also engages in several dance contests, in which the player must follow scrolling screen commands to press certain buttons in time to the music. CJ doesn't have b-boy moves, but he's a good dancer; his easy grace on-screen stands in comic contrast to the tense button-mashing of the player at the controls (Burrill 2006). When CJ is left unattended—when the player leaves him standing somewhere without pausing the game—he will eventually sing fragments of songs from the radio. Music, especially hip-hop, is integral to this avatar's character.

Unlike the previous GTA games, *San Andreas* requires players to alter the avatar's appearance in the course of quotidian gameworld life (see Chapter One). CJ is always a young black man, but his look changes substantially when he works out at a gym, eats too much fast food, gets a new hairstyle at a barbershop, or buys new clothes at various retail stores in the gameworld. Picking the wrong hairstyle, body type, or clothes for CJ can reduce a player's rating for "sex appeal" or "respect"; the player's taste may not match that of CJ's friends and girlfriends. Here *San Andreas* both makes fun of American consumerism and functions as a taste-making apparatus. The game cultivates a sense of what's cool and encourages players to experiment to see what choices will yield the best results. Players discover that different women prefer different body types and that a look that commands "respect" from gang members may not be what's attractive to the ladies. Similarly, the San Andreas radio stations are not value-neutral. As one player wrote, "The music, like other facets of the game (weapons, clothes, groupies), creates a level of connoisseurship in defining your character as a true avatar" (Drew Massey, Cambridge, Massachusetts, 22–26).

The songs on the *San Andreas* dial are unchanged from their actual-world versions, but DJ patter and station ads provide plenty of cues as to the coolness quotient of each station. If players tune in to K-DST, San Andreas's classic rock station, they hear a station ID that proclaims, "Anyone can play songs about abusing drugs and women, but we've been doing it successfully for 22 years." The DJ on Radio X, the alternative rock station, is a self-absorbed, disaffected slacker, a stereotypical privileged Gen-Xer; she complains about a "rap crisis," demonstrates her lameness by mangling the names of two rappers who are characters in the game, and suggests that being a gangster sounds like way too much work—she'd rather stare at candles all day. Choosing the house music on SFUR (San Fierro Underground Radio) means allying oneself with a flamboyant, arrogant German DJ: "I love you San Fierro! . . . Hug me, all of you! . . . I've not been this mashed since I was in India, communing with nature by hanging out with other people who had dropped out of work. Now we have a real record, coming at you from deep in the underground like a burst of magma!" On the modern hip-hop station, by contrast, an earnest and knowledgeable-sounding DJ promises only the freshest and most authentic gangsta rap—"N.W.A. expressing the reality of that West Coast life, man. . . . We get the music first, before anybody gets it. We locked to the streets." In this way *San Andreas* provides substantial guidance for players' sense of what constitutes good taste, authentic music, and a gangsta aesthetic—but it leaves them free to choose classic rock over classic hip-hop if that's what they prefer.

HEARING WITH CJ'S EARS

Today I had to take CJ to the barbershop—he'd been told to get a haircut and some appropriate gang colors, green for Grove Street. His friend Big Smoke took him to their old barber, or rather CJ drove Smoke's car (CJ always gets to drive). Smoke had his car radio tuned to classic funk, on Bounce FM, and the elderly black barber was listening to the classic soul station in his shop. There's a lot of old music in this game, more than I expected. And after hearing Charles Wright's "Express Yourself" in the barbershop, the samples in N.W.A.'s "Express Yourself" jumped right out at me from the gangsta rap station. (gameplay notes, May 9, 2006)

What kind of music would CJ listen to? It isn't hard to guess, given that the game shows us what his friends and siblings play in their cars and homes. Through characters like the local barber, we also learn what his parents' generation might have preferred. Other characters don't react to the music CJ chooses, but they do demonstrate their own tastes by example. A Chicano gang plays Cypress Hill. A young black DJ with a sound system in the back of her van plays the grrl-power rock band L7 for a beach party, but the van's radio is tuned to classic hip-hop. Each time I attempted a particular drive-by shooting mission, my car radio suddenly switched to the gangsta rap station with no input from me—an apparent programming glitch, but a cue to the player nevertheless. My survey respondents were very much aware of these connections. As one man wrote, "The game has a lot of socioeconomic themes, and the early '90s gangsta rap on Radio Los Santos frames a lot of the issues that the characters experience in a deeper way than the cut-scenes can. There's also—I think intentionally—a humorous bit of stereotyping when you steal a vehicle, in that the station seems be tuned to what the driver would likely be listening to; i.e., a pickup truck in the country will be tuned to KROSE" (Morgan, Virginia, 22–26).

The San Andreas radio stations encompass enough genres to appeal to the tastes of a wide range of players, and those genres are explicitly linked to the tastes of in-game characters. When CJ carjacks drivers in different areas, the victim's taste often matches the locale—country and rock out in the sticks, urban music in the inner city. But regardless of the genre, the radio seems to encourage players to listen with CJ's ears. The DJ patter on different stations falls along a spectrum from the intensely satirical to the relatively sincere, with the music CJ would take seriously receiving more respectful presentation. Dan Houser and Lazlow's scriptwriting invites listeners to mock the campy house DJ, to be amused and perplexed by the outlandish remarks of the country DJ, to be annoyed by the moody, cooler-than-thou alternative rock DJ, and to engage with the hip-hop stations that provide such good driving music and lessons on the early '90s West Coast rap canon. This radio system adheres to a fundamental game-design principle for representing the physical world: "The actual dimensions of the represented world are not dependent on their referent, reality, but on the capabilities and narrative goals of the characters. Gaps between ledges in the worlds of *Tomb Raider* are spaced exactly according to the abilities of Lara Croft and are not imitating the product of erosion or other natural causes" (Whalen 2004).

Players have their own opinions about the genres on the *San Andreas* dial, and most of them probably have previous experience with some of the songs. It is the contextualizing material that encourages players to listen as CJ would: the DJs, the ads, and the kinds of people who listen to each station in the gameworld. San Andreas DJs often discuss or directly address gang members and rappers, or at least that's what CJ hears. They may reproach or revile his social scene, but they acknowledge its existence and its currency; the classic hip-hop station promises "music from a time when your skills on the mic were more important than the size of your Gat," while the alternative rock DJ sighs, "Living in the suburbs is *so* much worse than the ghetto." CJ listens from the center of his own social world, as we all do, and hears the radio speak to him. As Jody Berland writes, "It is through [the DJ's] voice that the community hears itself constituted, through that voice that radio assumes authorship of the community, woven into itself through its jokes, its advertisements, its gossip, all represented, recurringly and powerfully, as the map of local life" (1994:185). *San Andreas* dramatizes the subjectivity of a particular listener as he encounters those radio voices.

Figure 2.1 suggests one way of thinking about "hearing with CJ's ears." It shows how the San Andreas stations tend to create a musical world marked by black/white, urban/rural, youth culture/older, and contemporary/classic dichotomies. Here my placement of the house/techno station, SFUR (San Fierro Underground Radio), merits further explanation. While "house" is something of a big-tent genre classification at this point, the vast majority of the names on SFUR's playlist are well-known African American artists, most with Chicago house credentials (e.g., Marshall Jefferson, Mr. Fingers, Frankie Knuckles, Joe Smooth, Maurice). But the Eurotrash DJ "Hans Oberlander" makes it clear that CJ would not hear SFUR as a black music station. For example, at one point the DJ suggests that an "ethnic track" reminds him of "backpacking and getting serious food poisoning. Ja, you really get to know the toilet early in the Third World." Like Radio X—the alternative rock station with the whiny (and implicitly white) female DJ—SFUR represents a youth culture unappealing to CJ. Thinking about the radio in this way might also explain what one player called "the glaring lack of a Spanish-language channel" in *San Andreas*. *Vice City's* Miami had such a station, as this man noted, and "it seems to be a big oversight that there isn't one for a fictional California" (anonymous male, 31–35). Indeed, two hits by Cypress Hill and Kid Frost provide the only nods to Chicano music in all of *San Andreas*. But this game depicts substantial tensions between black and Chicano communities, while also presenting a narrative thread in which CJ gradually becomes close friends with a Chicano character who is dating his sister. It would be difficult to script a radio station whose tone shifted to match this character development.

The value spectrum of the San Andreas stations is evident not only in scriptwriting but in casting choices: The country, house, and alternative rock DJs are played by professional voice actors in ultra-parody mode, while the hip-hop, contemporary soul, funk, and dub DJs are all famous representatives of those genres. Radio Los Santos is closest to the center of CJ's cultural world, and that station's DJ also provides the most direct bridge to the real world. He is Julio G, a real-life radio DJ who played groundbreaking West Coast gangsta rap on L.A.'s KDAY in the late 1980s. Julio G goes by his own name on San

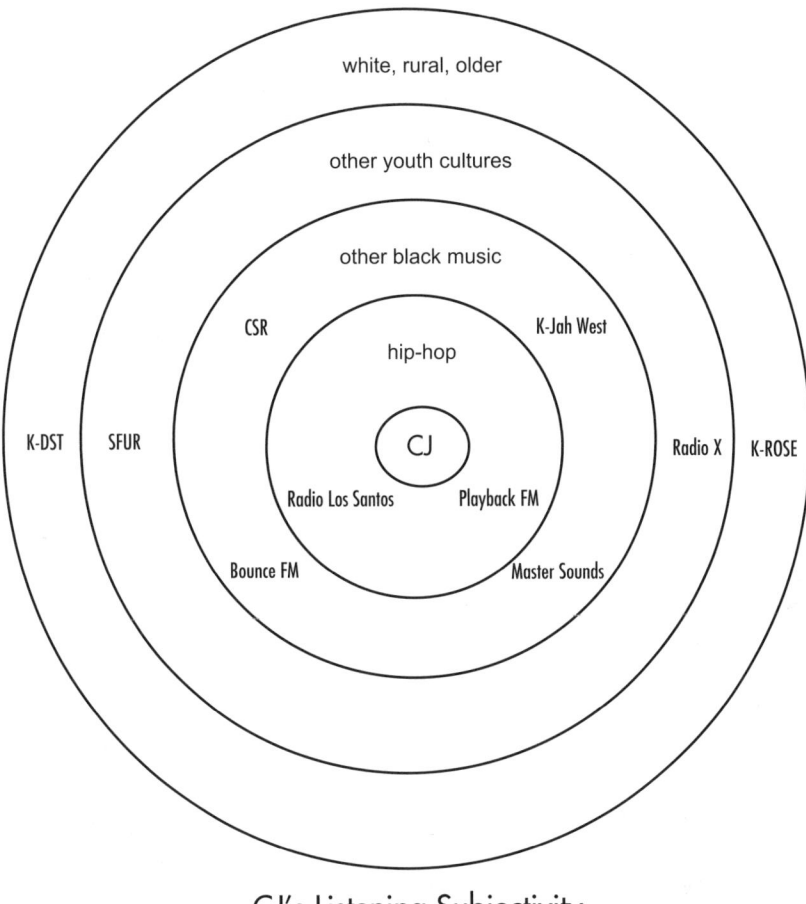

CJ's Listening Subjectivity

Figure 2.1
Hearing with CJ's ears. (*Diagram created by the author*)

Andreas's "modern hip-hop" station, bringing gangsta rap and G-funk to Los Santos. Meanwhile, Chuck D of Public Enemy is "Forth Right MC" on Playback FM; well known for his activism and political engagement, he lends considerable credibility and a dash of comedic self-righteousness to this "classic hip-hop" station. George Clinton goes by the name "The Funktipus" as host of the classic funk station. Michael Bivins of New Edition and Bell Biv Devoe heads up the contemporary soul station, while the legendary reggae production team Sly & Robbie speak for K-Jah West. Finally, the classic rock station K-DST presents a special case: The DJ is played by Axl Rose, who certainly qualifies as famous; but his script is tedious, his performance falls flat, and he presides over a list of painfully overplayed songs (e.g., Lynyrd Skynyrd's "Free Bird" and Eddie Money's "Two Tickets to Paradise"—CJ would be familiar with the K-DST songs, one imagines, but would not be a fan). All the DJs who are real-life musical artists can point to their own music somewhere on the San Andreas dial; presumably this was part of their contracts. Several players have suggested to me that Axl Rose's work on K-DST is the only possible explanation for the presence of a Guns N' Roses song on Radio X, the alternative rock station.

"Master Sounds," which plays mostly classic soul, does not have a famous soul artist at the microphone; the DJ is a fictional character voiced by black comedian Ricky Harris. However, "Master Sounds" has special authoritative status in San Andreas for other reasons. At least eight of its 18 songs have been sampled in songs that appear on the San Andreas hip-hop and contemporary soul stations (see Table 2.2). Moreover, all 18 have been sampled by San Andreas soundtrack artists at some point in their careers.[7] In the two cases where the hip-hop song and the classic source share a title—"Express Yourself" by Charles Wright/N.W.A. and "I Know You Got Soul" by Bobby Byrd/Eric B. & Rakim—the connection would be hard to miss for even a casual listener. A station ID drives the point home:

MALE VOICE: Master Sounds FM.
FEMALE VOICE: The only place you can hear the original before it got sampled and some jive turkey started chanting over it and not paying royalties— hey, where's my bread, man?
MALE VOICE: When you're done stealing beats, gangstas, listen to Master Sounds.

This station is explicitly presented as a hip-hop history lesson. Its songs come straight out of the "mom-and-pop crates" described by Joseph Schloss: "a certain core of well-known records, generally those that were popular with urban African American listeners in the 1970s" (2004:82). (Early hip-hop artists often got their starting material from their parents' vinyl collections; "mom-and-pop crates" is shorthand for a record collection that highlights these canonical 1970s releases.)

Many of the songs on the classic funk station "Bounce FM" also have iconic status as sample sources, though the presentation there is more party-oriented than didactic—freaky, not teachy. A comprehensive review of the *San Andreas* soundtrack for the gamer website *IGN.com* listed over 40 artists who have sampled the Ohio Players' "Funky Worm" (1972), among them nearly all of San Andreas's hip-hop artists (Spence D. 2004). This song's warbly, squelchy, high-pitched synthesizer sound—produced by a mono synth with portamento and resonance effects—became a signature sound of G-funk hip-hop; thus "Funky Worm" now has West Coast/G-funk associations for certain listeners, though the Players were in fact from Dayton, Ohio. Other widely sampled songs on "Bounce FM" with special place-resonance for San Andreas include Kool & the Gang's "Hollywood Swingin'" (1973) and Ronnie Hudson's L.A. car-culture anthem "West Coast Poplock" (1982). The latter's lyric reference to

Table 2.2 CROSS-STATION SAMPLING FROM THE
"MASTER SOUNDS" PLAYLIST

Charles Wright, "Express Yourself" (1970)
=> Radio Los Santos / N.W.A., "Express Yourself" (1989)
Bobby Byrd "I Know You Got Soul" (1971)
=> Playback FM / Eric B. & Rakim, "I Know You Got Soul" (1987)
Bobby Byrd, "Hot Pants" (1971)
=> CSR / Johnny Gill, "Rub You the Right Way" (1990)
Bob James, "Nautilus" (1974)
=> Playback FM / Slick Rick, "Children's Story" (1988)
Maceo & the Macks, "Soul Power 74" (1974)
=> Playback FM / Spoonie Gee, "The Godfather" (1987)
Lyn Collins, "Think (About It)" (1972)
=> Playback FM / Slick Rick, "Children's Story" (1988)
=> Playback FM / Rob Base & DJ EZ Rock, "It Takes Two" (1988)
=> CSR /Boyz II Men, "Motownphilly" (1991)
The J.B.'s, "Grunt" (1972)
=> Playback FM / Public Enemy, "Rebel Without A Pause" (1988)
=> Playback FM / Eric B. & Rakim, "I Know You Got Soul" (1987)
James Brown, "The Payback" (1974)
=> CSR / En Vogue, "My Lovin' (Never Gonna Get It)" (1992)
=> Radio Los Santos / Da Lench Mob, "Guerillas in tha Mist" (1992)

the "city of Compton" was featured by N.W.A. in the seminal West Coast track "Straight Outta Compton" (1988), which would have to be numbered among the state of San Andreas's foundational cultural texts.[8]

References to Compton and South Central L.A. are frequent on Radio Los Santos, home of "modern hip-hop." This station's version of what's "modern" is actually what's local. Some of the songs are as old as those on Playback FM, but they are all by West Coast artists. Members of N.W.A. had a hand in 11 of the station's 16 tracks, and contributions by groups like Compton's Most Wanted and Cypress Hill further establish the connection to L.A.[9] Radio Los Santos constantly reinforces the relationship between Compton and "Ganton," CJ's home neighborhood. For example, as the player drives through foggy streets, DJ Julio G links Da Lench Mob's "Guerillas in tha Mist" (1992)—which depicts South Central as a literal jungle—to the current Los Santos weather: "Today coming to work I came through this thick, thick fog, for like 10 minutes, and it was crazy, man. And so what I did, I went to the back seat, got my Lench Mob CD, and I put this one in. I was thinking about y'all, so I'm gonna play it again for all y'all. 'Guerillas in tha Mist,' it's the Lench Mob, y'all."

This dense web of interconnected songs writes the history and shapes the cultural landscape of the state of San Andreas, always reaching beyond the borders of the gameworld to make connections with real-life songs, cities, and musical canons.[10] While the radio puts forward "classics" in many genres, it is African American music that receives the richest, most thorough, and most historically aware treatment. In San Andreas, "Rap music is a school system itself, and one of the best school systems that we have," as Ice Cube has said (Ice Cube and Davis 1992:189). Hearing through CJ's ears means hearing those musical connections and their associated cultural history as a part of everyday life, as the background music for barbershops, house parties, dance clubs—and for drive-by shootings.

PLACE, TASTE, AND CHARACTER

What do players make of San Andreas's musical soundscape and CJ's relationship to it? One answer is that players who normally listen to alternative rock start listening to rap, because that's what CJ would do. They decide that inhabiting a character means acquiring his tastes, and they follow the game's cues in making the consumer choices of a stereotypical young black gang member. When I asked players, "Do you think music has any effect on how you relate to your character or other characters in the game?" virtually all of the positive answers referred to rap and *San Andreas* (even though all these respondents had played other GTA games and referred to them in responses to other questions).

> Absolutely, all the hip hop/rap type of music, even though I'm not a fan of it, help[s] me to become more immersed in the game. (anonymous male, 31–35)

> i wouldn't usually listen to Rap, and only do ingame because CJ would. (anonymous male, teens)

> Yes. The rap music in SA seems to fit the best to the game's situations and I feel that it is the most appropriate music to play while performing anti-social behaviour. (Scott, Canada, 22–26)

Another respondent reported that while he personally doesn't listen according to CJ's perceived tastes, his brother does: "i had on the rock station, and he changed it and said 'my nigga wouldnt listen to that'" (follow-up email correspondence).

Players need not rely on stereotypes to associate CJ with rap; along with all the gangsta-oriented talk on the radio, the game provides cut-scene dialogue that establishes CJ's preferences. In an early mission CJ tells another character, the aspiring rapper OG Loc, "I really ain't into rap no more, I'm more into that hardcore gangsta shit." This suggests that even Playback FM might not be quite to CJ's taste—which would explain why Chuck D's script on that station comes across as a bit preachy, focused on asserting what constitutes "classic hip-hop" rather than playing what's edgy, local, and true to CJ's experience.

Meanwhile, the Radio Los Santos playlist starts with 2Pac's "I Don't Give a Fuck" (1991), which Julio G identifies as brand new, adding that he is bound to get a fine for the foul language. This song and the others on Radio Los Santos virtually write the script of *San Andreas,* with their focus on West Coast locations, hostile cops, black-on-black crime, and the inexorable everyday grind of violence: "Just another day in the life," says 2Pac.

But while many players associate gangsta rap with CJ or his activities— they listen to Radio Los Santos because that's what CJ would do—other perspectives were just as common among my respondents. For one thing, since *San Andreas* provides only about 10 hours of music for many tens of hours of gameplay, it's hard to imagine someone sticking to any one station for long. As a Dutch survey respondent explained, "In the beginning of GTA San Andreas i thought, CJ is a gangster, so i listened to gangster station Radio Los Santos. After a while, i had heard all songs on that station so many times, that i wanted another station and i searched more for music i like myself, and not music i liked and which fitted the game" (anonymous male, teens). Several respondents suggested that they identify more closely with an avatar if he seems to share their own tastes. Rather than listening with the avatar's ears, they preferred to think of the avatar as listening with theirs.

> Since the music I as a player [am] listening to is the same as the music the in-game character is listening to, it makes me feel a bit closer to him, whereas a more traditional soundtrack would create a layer of distance. (David Newgreen, Melbourne, Australia, 18–21)

> [I] relate to CJ better because there's more rock stuff. If it was purely rap, which I don't mind, but it would get boring, it would be a less involving game. (anonymous female, United Kingdom, 18–21)

> I think the radio helps me to identify more with the character that I am controlling in the game. I am more convinced that he is like me if I can not only control most of his actions, but also what kind of music he listens to. If he stole a car and Creed was on the radio and he didn't change it, that would serve to ruin the illusion. (Kevin McCoy, Michigan, 22–26)

Finally, whether or not they associated some of GTA's music with particular game characters, nearly all the survey respondents connected the radio with the gameworld's historical setting, locations, and their own emotional states during gameplay.

Since the vast majority of my respondents had played *Vice City* before playing *San Andreas,* they were accustomed to a set of radio stations that represented the gameworld's era rather than the avatar's personality. But instead of representing 1992 as a stand-alone cultural moment or the West Coast as a culturally diverse region, the *San Andreas* radio sets up the oppositions sketched in Figure 2.1: black/white, urban/rural, young/old, classic/contemporary. These distinctions correspond closely to different parts of the gameworld, so players tend to match the stations to particular *San Andreas* locales and vehicles. In so doing, they follow the lead of the game designers, who made a stolen truck in the countryside most likely to be tuned to classic country or classic rock.

> Country music is more appropriate for a stolen pickup truck, or hardcore rap for a lowrider. And some of the more relaxing stations are better suited for long drives in the country, while others may be better suited for, y'know, doing a drive-by. (Joe, Little Rock, Arkansas, 27–30)

> I tend to enjoy listening to rap in the city and the rock stations while driving through the country. In the country, the country-western station also feels appropriate and I listened to it while driving certain vehicles such as campers and tractors. (Scott, Canada, 22–26)

> Radio LS is my favorite when I'm in LS, I usually listen to K-yah when in San Fierro, KDST is my desert radio station and when in Las Venturas city I particularly like listening to mastersounds. (anonymous male, The Netherlands, 18–21)

Other players focused on the connection between music and mood: for example, linking faster, harder, more aggressive-sounding tracks to violent activities; choosing music that seemed to help them complete difficult missions; or deliberately creating surreal contrasts by mismatching music and activity. In my own gameplay I discovered that the dub/reggae station helped me relax during high-speed car chases—other stations offered equally mellow music, but K-Jah West reminded me of the island music of my childhood in Key West, Florida. Survey respondents often reported making similar performance-enhancing choices based on their own tastes and histories. A few players who chose to listen to music from their own collections during gameplay described an increasingly strong link between particular music and particular game activities. As one survey respondent wrote, "What I did to myself, unthinkingly, was associate the sounds of Cannibal Ox with the virtual act of dangerous driving, shooting guns, and so on. So, if I was driving a friend's car, and he had put a Cannibal Ox song on the mix tape he had left in the player, I would experience what I can best describe as a game fugue; I would have the ghost of an impulse to drive the car as if I was playing the game" (anonymous male, 31–35). In some versions of the game, players can create custom mp3 playlists so that they hear their own music amid the game's radio advertisements and news bulletins—an experience that likely fosters the development of these associations. (However, my research—and Rockstar's investment in licensed content—suggests that most players thoroughly explore the existing stations first.)

Because *San Andreas* creates a sense of what constitutes appropriate music for particular places and activities, players are in a position either to pursue those associations or to subvert them. As a survey respondent wrote,

> I choose my stations based upon what I plan on doing. If I'm thinking about going on a crazy car chase, then I'll choose the appropriate high-energy station. If all I want to do is drive around and admire the scenery, I'll switch over to whatever relaxing yet interesting station there is. Of course, I could always do the opposites; there's great control given to the player, since we can change the mood of our given actions however we'd like. (Robert Salazar, Texas, 18–21)

In conversation and survey responses, players often brought up these possibilities when discussing drive-by shooting missions—a hallmark of the *San Andreas* driving experience. One player suggested that the radio "gives a much higher level of immersion versus other games, it's nice to do drivebys while listening to Dr Dre sing 'Nuttin but a G thing'" (anonymous male, 31–35). Another reported, "Sometimes I like to use the country station on San Andreas while doing a drive-by because it's so surreal" (Mike Berezowsky, Edmonton, Canada, 27–30). These players know what kind of music is supposed to go with mass-media depictions of ghetto violence; they enjoy being able to choose whether to comply with those norms or to create a sense of cognitive dissonance.[11]

This is not to say that all players are inclined to critique these stereotypes. Some use Radio Los Santos for drive-bys because it strikes them as the realistic thing to do. But *San Andreas* encourages experimentation with other musical options. As players investigate the stations on the *San Andreas* radio dial—with all their potential links to mood, character, activities, and locale—they develop a relationship with unfamiliar genres and artists that often leads to real-life purchases or downloads. My survey asked: "Has the GTA radio changed any of your listening/music-purchasing habits? Has it increased your interest in any particular artist or genre?" Over 65 percent of respondents gave affirmative replies.

> I downloaded most of the game's soundtrack after completing the game. I also feel connected to the game and the game's play when I listen to certain songs from the game during everyday activity. In particular, I have begun listening to a great deal of "gangsta rap" such as Cyprus Hill, Ice Cube, and Easy E [artist names *sic*]. I have also sought out music produced by many of the bands whose music was present in the game (including Pearl Jam, Tom Petty, etc). (Scott, Canada, 22–26)

> It's probably broadened my appreciation of both country music and reggae. (Joe, Little Rock, Arkansas, 27–30)

> Although I'd enjoyed in passing some of the '90's gangsta rap, it wasn't until I really got to know the songs in the game that I started tracking down the albums from

N.W.A. and Ice Cube and the like, which of course led to finding more contemporary political hip-hop. I think the repetition of the songs also made my ears more attuned to hearing the rapped lyrics more easily. (Morgan, Virginia, 22–26)

Being exposed to "classic hip-hop" has increased my awareness of the genre, and I'm giving the buyer at my local record store fits trying to buy ancient hip-hop on vinyl. (anonymous male, 31–35)

I was googling the search terms "grand theft auto radio" when I stumbled across this [survey] page. I was looking for a torrent file to download the audio so that I can listen to it in my car. (Ryan, 18–21)

I have also discovered evidence of GTA-driven purchases in many Amazon.com user reviews of particular albums, particularly those associated with Radio Los Santos and Playback FM. Wikipedia entries and other general-information websites on the musical artists featured in *Vice City* and *San Andreas* often mention that a particular song can be found in the games or on their spin-off soundtrack box sets. GTA's playlists have led players to seek out reissues of music from their childhoods and to investigate the historical roots of country, rap, reggae, and rock.

Here it is important to remember that the licensed music on the radio constitutes the only direct product placement in *San Andreas*. While the gameworld is rife with billboards, retail stores, radio advertisements, cars, and clothes that parody actual products—from the Urban Outfitters knockoff "Sub Urban" to an antidepressant called "Grin"—only musical artists and the DJ Julio G are directly imported from the actual world and appear under their real names. The *San Andreas* instruction booklet/tourist guide contains advertisements for all kinds of fictional in-game businesses, but the only real ads are for music. Amid the fake ads for "Caligula's Casino" and "Ammu-Nation" ("for all your daily firearm needs"), full-page ads from fictional record stores reproduce real album covers under the heading "The hottest hits on sale this weekend": 2Pac's *2pacalypse Now* (1991), Boyz II Men's *Cooleyhighharmony* (1993), *Appetite for Destruction* by Guns N' Roses (1990), and Public Enemy's *It Takes a Nation of Millions to Hold Us Back* (1988), among others. At the end of the tourist-guide section, just before the game credits, Capitol Records gets a two-page ad spread with no contextualizing references to the gameworld. The covers of Ice Cube's *Death Certificate* (1991), Eazy-E's *Eazy-Duz-It* (1988), and N.W.A.'s *Straight Outta Compton* (1988) appear opposite coming-in-2004 releases by Houston, Chingy, and Don Yute.

With its ability to deliver multimedia content to millions of players around the world, Rockstar Games is an attractive partner for record companies. Given that many Rockstar staff members have professional backgrounds in music promotion, label management, and music video production, the company is in a position to make the most of this synergistic potential. Its gameworld playlists must be analyzed with this in mind. However, it should be noted that many players are aware of the licensing/marketing issue and have praised Rockstar's choices in this arena, in terms of both playlists and editing. As a survey respondent wrote,

> The main difference [between music in GTA and other games]: They openly drop the F bomb . . . Nothing kills me more than dead air in the middle of a song because the developer wanted to retain their T for Teen rating while still including some hot, edgy song on their playlist. The music is also mostly songs you don't normally hear. An EA [Electronic Arts] sports title, for example, will include whatever is popular and current. The radio in GTA includes tracks which are both relevant to the era, but tracks and bands that you don't normally hear or do not normally get any air play. (Matthew, Keokuk, Iowa, 22–26)

PLAYING ALONG WITH A HIP-HOP INDUSTRY STORY

As I conducted my virtual fieldwork in Vice City and the state of San Andreas, my sense of the musical cultures of these simulacra societies emerged not only from the songs on the radio but from the gameworlds' representation of the commercial music industry. GTA's designers successfully market music to players while implicitly critiquing the mass-media star system and its promotional machinery. Rock-star, rapper, and band-manager characters interact closely with Tommy Vercetti and CJ, and some of them are interviewed or discussed by in-game radio DJs—a practice that blurs the line between satirical and sincere music promotion on the GTA radio. While the metal band Lovefist provides Spinal Tap–style comic interest in *Vice City,* in *San Andreas* a more complicated music-industry story is woven through the narrative: the tale of the rappers OG Loc and Madd Dogg.

As Soraya Murray has observed, *San Andreas* is set just at the leading edge of the bling-bling, hypercommercialized rap era that has been excoriated not only by the moral-panic crowd but by many ardent hip-hop supporters (2005:96). I have already suggested that the contrast between Playback FM and Radio Los Santos underscores and historicizes this transitional moment, shoring up CJ's distinction between ordinary rap and "hardcore gangsta." But gangsta rap itself has two faces in *San Andreas,* personified by the ambitious, inept Jeffrey Cross, who has taken the name of "OG Loc" (voiced by actor Jonathan "Jas" Anderson), and the successful rapper Madd Dogg (voiced by Ice T, whose music is conspicuously absent from the Radio Los Santos playlist).

Jeffrey/OG Loc is a young man from CJ's neighborhood, probably a teenager. We first meet him when he is released from a short prison term, which he sought out for the express purpose of increasing his street cred as a rapper. His prison comportment has been less than gangsta, however—as CJ's brother tells another character, "You know Jeffrey's been somebody's bitch for the past three weeks, right?" In cut-scenes, CJ and his fellow workaday gang members are half-amused and half-disgusted by Loc's showy gold chains, tattoos, and awkward attempts at ghetto slang; they often forget to use his rapper name and revert to calling him "Jeffrey." Before long Loc enlists CJ to help him find and kill his prison boyfriend, thereby eliminating the risks that kind of sexual history would pose to his rap career. CJ has his own reasons for doing jobs for Loc: His older brother, Sweet, constantly questions CJ's commitment to the Grove Street Families, and these missions provide a means to regain Sweet's trust.

OG Loc practices his rapping while he works as a "sanitation technician" at a fast-food chain called Burger Shot, where he has to wear a humiliating burger-shaped hat and do unpleasant jobs like cleaning the deep fryer. He persuades CJ to steal a sound system so that he can throw a party to promote his first record. At the party, CJ and his friends wince at Loc's terrible rhymes. Shortly thereafter, CJ visits Loc as he cleans the Burger Shot men's room:

[LOC is mopping the floor in his burger hat.]

LOC: Hey hey hey, it's OG Loc, homie, and I'm gonna kick a little something like this. Hey yo, when I come through up in the place, you don't want me to come with a gun in your face. I spit it harder than anyone in the world could do . . .

CJ: Hey, dude, you ever thought about getting a writer to help you out with this shit?

LOC: Yeah, I did. But who, homie, who?

CJ: Shit, I ain't in the rap game. You know that ain't my thing. But, um, we have to think about something. Man!

LOC: How 'bout if I get somebody to write something for me, only they didn't know about it? . . . I think I just might have found a ghost writer. I'll become the reciter, all nighter, all righter—Madd Dogg's rhyme book, from his home in the hills.

CJ: Madd Dogg's rhyme book?!

LOC: Man, you said you'd help, Carl. Come on, man. I'm hot, like fire all nighter all righter, when I kick it—

CJ: Hey! I'll do anything, homie, I swear that. 'kay?

CJ sets off to steal Madd Dogg's book of rhymes for Loc, in a storyline that reminds players of the written-out literary qualities of rap lyrics and the aesthetic distinctions between good and bad flow. The mission takes place in a successful rapper's gigantic, near-empty mansion in a boring nouveau-riche neighborhood, far from the active street life of Ganton. When CJ sneaks into the barely furnished house, Madd Dogg is sitting alone playing video games while armed guards patrol the markers of his financial achievements.[12] This is the lifestyle that signifies commercial success—"I write my rhyme as I cool in my mansion," as Rakim raps on Playback FM in "I Know You Got Soul" (1987). It certainly looks a lot better than OG Loc's job at Burger Shot. Still, CJ's life as a real gang member trumps them both with its independence, authenticity, and community connections.

No one can stand OG Loc's posturing, but he becomes famous anyway. After CJ gives him the rhyme book and drives Madd Dogg's manager off a pier, OG Loc begins to get mentioned on the radio. Soon we hear him interviewed by Lazlow on *Entertaining America*, a mainstream talk-radio show—clearly a big break.

OG LOC: I'm ice cold, bitch. Don't let me dump on you, G. I'm the streets, man, I am *gangsta*. I'm taking rap in a whole new direction . . .

LAZLOW: Why do you rappers get so worked up? You're rich! You've won! Stop shooting at each other. You know, and you keep saying "I'm from the streets." You know what, dude, everyone has a street in front of their house. That doesn't make you cool.

OG LOC: Oh, we've got a comedian, huh. You got straps, bitch, huh, you down? You punk, trick, busta, fool!

LAZLOW: Look, I don't know what you said, but here, this oughtta calm you down, I brought you some malt liquor.

OG LOC: You's a busta fool. Lucky I don't hang you out the window or turn you out, 'cause I'm also a pimp. Including dudes, I'll pimp anything, y'hear me?

LAZLOW: Oh! Dude, I hear you loud and clear . . . Please, don't shoot me, homie!

OG LOC: Relax fool, no one's getting dumped on. I'm a warrior poet, I tell a cautionary tale about life on the streets, you know?

LAZLOW: Only too well. That was OG Loc—hey, man, it's been a real pleasure. Straight. Yeah. Good luck with the music . . .

ANNOUNCER: So *that's* how *you're* going to be entertained.

As OG Loc becomes more and more of a sellout gangsta stereotype, subjecting himself to condescending and racist media treatment, Madd Dogg's career is destroyed. When CJ next encounters him, Madd Dogg is a burnt-out drug addict, about to commit suicide by jumping off a building. (In accordance with GTA's signature dark humor, the player has to rescue him by maneuvering a truck to catch him when he falls.) Guilt-stricken, CJ sends Madd Dogg to rehab and offers to become his manager. They return to Los Santos together, recover the mansion from drug dealers, and commence recording some new tracks. The in-house recording studio is detailed and realistic; one player was certain he recognized Yamaha NS10 studio monitor speakers and a Tascam DA-88 digital multitrack recorder.[13]

By this time CJ has completed missions in the rural countryside, San Fierro/San Francisco, and Las Venturas/Las Vegas. He has developed excellent business relationships in several industries and has demonstrated his ability to move in many social circles, rather than being mired in the mean streets of Ganton. When he does a favor for a government agent, earning Sweet an early release from prison, CJ is excited to share his accomplishments with his brother. But Sweet is apparently as disappointed in CJ as ever.

CJ: Hey, man, we're off to our new spot. We got a mansion, Sweet! We been puttin' in work, and shit is goin' *well*. We got a stake in a casino, we got some insane shit in [San] Fierro, we gettin' into the rap game! Hey, man, let me get you some new clothes, come on! . . .

SWEET: You never did get it, did you, Carl? I need to go check on things in the hood. Man, that's the problem. You're always a perpetrator, runnin' from what's *real* . . .

CJ: Man, what the hood done for me? Always draggin' *me* down. Ever since I *got* out the hood, shit been crackin'! That's everybody dream, to get out the hood . . . A'right man. You hard. I'm gonna *show* you what's going on in the hood . . .

The player must drive Sweet to Ganton, where the brothers encounter a drug addict trying to sell them their own mother's blender. CJ says, "Look like baseheads have took over the spot. Let's go home." Sweet's reply is swift and furious: "This *is* home, man . . . You was born in there!"

CJ's apparent freedom is constrained by this moral pressure from his brother, here and throughout the game. Rather than leaving Ganton behind, he has to resume a leadership role in the Grove Street Families and help clean up the neighborhood (eventually killing an old friend who has become a drug kingpin). Madd Dogg gets a new contract from a white man at "Blastin' Fools Records." By the end of the game he is back on track for the life described by Too $hort on Radio Los Santos, where "Even though they put us down and call us animals /We make real big banks and buy brand new clothes /Drive fancy cars, make love to stars /Never really saying just who we are" (1990). But CJ's future is unclear. Even after he defeats the game's villains and the *San Andreas* credits roll, the player can continue with free-form gameplay. What's obvious is that CJ is compelled to stay committed to Ganton. Despite his explorations of the whole state of San Andreas, in the end his is the story told by Compton's Most Wanted in "Hood Took Me Under": "A nigga can't escape the gangs and the dope . . . /Cause when you grow up in the hood, you gots to claim a set. / Yeah, it's not that you want to but you have to" (1992). Sincerely or ironically, this dramatic conclusion has a strong antidrug and "family values" message. It brings to mind the plots of blaxploitation films, with their "peculiar mix of fantasy and pragmatism, whereby criminal behavior is at once commended and deplored" (Demers 2003:42).

In tracing the intersecting careers of OG Loc, Madd Dogg, and CJ, *San Andreas* guides players through the foundational narratives and central dramatic conflicts of gangsta rap. As Eithne Quinn writes, this genre was created by artists who were "at pains to expose and critically engage [gangsta rap's] own commercial impetus and commodified status" (2005:5). Commercialism and the co-option of stereotypes could be seen as just part of being a good hustler, making a living by any means necessary. OG Loc's methodical steps to industry success—spending a little time in prison, proving his "hardness" with violence, appropriating another MC's rhymes, and pimping himself out to the mainstream media—present a merciless portrait of that kind of hustle. bell hooks has associated this phenomenon with what she calls the "shift in class values" in African American life that came with "the idea that money is the primary marker of individual success, not how one acquires money" (2004:19). She critiques mainstream hip-hop culture as "just a black minstrel show—an imitation of dominator desire, not a rearticulation, not a radical alternative" (152).

In *San Andreas,* too, money is a key measure of success. Players earn dollars, not points. But at the same time, CJ's relationships with OG Loc, Madd Dogg, and Sweet pull him in different directions. As Tricia Rose notes, "The hood is not a generic designation" (1994b:10)—or, to use Sweet's terms, "You was born in there." CJ must negotiate the conflicting demands of financial ambition, loyalty to friends and family, and dedication to authentic street life in a particular neighborhood, while also dealing with a racist, corrupt police force and the predations of violent drug dealers and rival gangs. This is a story told many times in rap lyrics. Both sensationalist and didactic, it presents "an utterly complete index of social ills" (Wheeler 1991:205).[14]

In creating *San Andreas,* Rockstar's designers not only drew on their own experience in the music business but also hired writers, consultants, and voice actors with strong hip-hop affiliations. DJ Pooh—a frequent Ice Cube collaborator, former producer for L.A.'s Da Lench Mob, and screenwriter for the classic ghetto comedy *Friday*—is one of only three credited writers for *San Andreas.* West Coast hip-hop photographer/music video director Estevan Oriol contributed to the visual design of the game, as did Mexican American tattoo-artist-to-the-stars Mister Cartoon. Setting aside the in-game radio DJs, hip-hop artists who play named characters in the *San Andreas* cast include Ice T, Yo-Yo, MC Eiht, The Game, and Kid Frost. CJ himself is played by an aspiring West Coast rapper, Chris Bellard (Young Maylay), who released an album based on his *San Andreas* experience (Young Maylay 2005). This extensive hip-hop pedigree gives *San Andreas* a consistent aesthetic feel that ties together the game's visual design, characters, dialogue, cut-scene editing, musical content, and narrative—all of which sample from other media sources (see Chapter One). Like hip-hop, *San Andreas* is driven by Signifyin(g) practices, always juggling earnestness and irony (Gates 1988, Wheeler 1991). The game shares what Rose has described as the "fundamentally literate and deeply technological" character of rap music, whose artists "reshuffle known cultural formulas and themes" to generate new and compelling narratives (1994a:95, 88).

The *San Andreas* radio stations and music-industry stories encourage players to draw conclusions about CJ's motivations and cultural history as they relate to the nature of authentic hip-hop. Canonical rap tracks that closely match the game's time and place give *San Andreas* intertextual richness and historical depth. One survey respondent wrote, "I find that it is one of those things that really made me feel like it WAS that time. Having Dr. Dre blasting while you watch black people being chased by white cops shows a very strong message" (anonymous male, teens). At the same time, the radio helps preserve the delicate balance of gritty realism and comic camp that has made *Grand Theft Auto* so successful and so controversial. As another teenage respondent explained,

> the radio makes the game feel overall more realistic but with the funny and cartoon-esque dialogue of the radio hosts it enforces the unrealistic nature of the game. in a sense, u feel that u can really do these things for one moment, then u realize its impossible (only when u think about [it], before that i can escape from reality in the game) (Ben, Ontario, Canada)

These young men were small children in 1992, the year represented in *San Andreas*. In 2004, the year of the game's release, they were playing in defiance of its age rating. They're exactly the kind of impressionable youths that critics say should not be immersed in GTA's virtual worlds, called "virtual reality murder simulators" by activist/attorney Jack Thompson in a 2006 lawsuit.[15] But my research suggests that while *San Andreas* does include violent content, it is also communicating some complex historical, political, and ethical lessons about hip-hop, consumerism, and the life circumstances of characters who find themselves choosing between deep fryers, criminal pursuits, and musical entrepreneurship. OG Loc's frustration with his dead-end Burger Shot job is palpable, while Madd Dogg's career is subject to the whims of a fickle, exploitative music industry. Such stories are enshrined in early gangsta rap—recall Ice Cube's "A Bird in the Hand," which describes a McDonald's job in withering terms and asks, "Do I gotta go sell me a whole lotta crack /For decent shelter and clothes on my back?" (1991). Stories like these do indeed "show a very strong message."

To the extent that *San Andreas* is historical, it adheres to Hayden White's dictum that "every historical narrative has as its latent or manifest purpose the desire to moralize the events of which it treats" (1980:18). This is particularly apparent in the game's re-creation of the 1992 L.A. riots, sparked by the acquittal of a corrupt Los Santos police officer. (Officer Tenpenny, a black cop voiced by Samuel L. Jackson, harasses CJ throughout the game; he bears some resemblance to the nefarious Officer Coffey from *Boyz N the Hood*.) As players navigate a Los Santos filled with looters, police, and raging fires, they get a history lesson not taught in most high schools. Like the real-life songs on the radio, this real-life event has been contextualized so that players experience it from CJ's perspective. They don't learn about the traumatic experiences of L.A.'s Korean Americans, for instance, nor do they hear the L.A.P.D.'s side of the story. But while acknowledging the limited nature of this representation, I still want to suggest that it might be an eye-opening experience to go through a reenactment of the riots as a young man whose own neighborhood is burning (cf. Murray 1997:170).

Like many of its hip-hop antecedents, *San Andreas* both romanticizes and essentializes "the ghetto." At the same time, its storylines vividly depict and critique the commodification of blackness that has made "fictive accounts of underclass black life . . . more lauded, more marketable, than other visions because mainstream conservative white audiences desire these images" (hooks 1994:152). The game thereby follows Ice Cube's prescription for rap: that it should "hold up a mirror," so that "you see yourself for who you are" (Ice Cube and Davis 1992:178). But while Ice Cube's mirror was meant to confront black youth with "the things going on in the black community" (ibid.), *San Andreas*'s mirror reflects a broader culture of consumerism, glorified violence, media sensationalism, and racial stereotypes. As art historian Soraya Murray has noted, the game "has taken the poor black male body, which is encoded as a human stain on the fabric of a squeaky-clean American dream of opportunity, and pushed it into the center of our attention. . . . That shell upon which so many negative associations has been projected becomes a mirror for a thorny cluster of societal relations" (2005:80). *San Andreas* could hardly be credited with offering bell hooks's wished-for "radical alternative" in representing black life (2004:152), but it does communicate the idea that actively critiquing media representations is both politically important and endlessly entertaining.

Nonlinear "sandbox" games like those in the *Grand Theft Auto* series aim to create an open-ended, experimental gameplay experience, but their technical limitations inevitably impose constraints on players. *San Andreas* gives that play of freedom and constraint cultural specificity and logical force by requiring players to occupy a particular avatar, a young black man with a detailed personal history. Moving with CJ's body, interacting with CJ's friends, living in CJ's neighborhood, and hearing with CJ's ears all serve to suture the state of San Andreas to hip-hop culture, a well-established arena for staging conflicts between individual expressivity and oppressive restrictions. As Rose writes, "Themes in rap and graffiti articulated free play and unchecked public displays, and yet the settings for these expressions always suggested existing confinement" (1994b:84)—a quality that aligns hip-hop with nineteenth-century blackface performance (Lott 1993:234, Lhamon 1998:16). *San Andreas* dramatizes that counterpoint of free play and societal confinement in extreme terms, celebrating hip-hop's opportunistic creativity but also undermining the "straight outta Ganton" myth of ghetto authenticity through broad parody.

By encouraging players to experiment with the possibilities and the confining borders of CJ's world, *San Andreas* invites them to interpret its controversial content on their own terms and to investigate their own complicity with the stereotypes that govern much of the social life of the gameworld. As the Kant-quoting country DJ Mary-Beth Maybell tells them, they can choose to view themselves as "passive recipients of perception"—manipulated by advertisers, doomed to reenact video game violence in real life—or as "active originators of experience," equipped to appreciate irony. Cultivating this sense of in-the-know interpretive freedom has been good for business at Rockstar Games. What player would choose to identify as a passive dupe of the culture industry? But this does not diminish the fact that the makers of *Grand Theft Auto* acknowledge and capitalize on their players' creative agency, musical sensibilities, and past experiences with popular media. By jacking the dial—appropriating commercial radio to broadcast new messages—these games show how popular music transmits social history and how players' musical choices can animate and enrich their exploration of virtual worlds.

PART TWO

Playing Along with Guitar Hero *and* Rock Band

CHAPTER 3
How Musical Is *Guitar Hero?*

The *Guitar Hero* and *Rock Band* video games are designed to put players in the virtual shoes of live rock-concert performers. Musicians record a song, game company staff transcribe it into the game's notation system, and players contribute their physical engagement and performance adrenaline to reconstitute the recordings as new performance events. Gameplay involves translating the on-screen notation back into music by pressing buttons on a plastic guitar or striking the pads on a simulated drum kit. When players follow the notation accurately, the prerecorded songs play through the television speakers. Meanwhile, a virtual crowd supplied by the game software cheers players' successes and heaps scorn on their failures. A flesh-and-blood audience might also be watching in the living room, bar, or tournament hall where the player is performing. Some of these performances are uploaded to YouTube, where millions of people can watch and comment at their leisure.

In 1970, Roland Barthes suggested there are two kinds of music: "the music one listens to" and "the music one plays." He dubbed the latter "musica practica" and wrote lovingly of the experience it fostered: "The body controls, conducts, coordinates, having itself to transcribe what it reads, making sound and meaning" (1977 [1970]:149). Barthes also asserted that musica practica had disappeared, replaced by listening; amateur musicians were "no longer anywhere to be found," or at least no longer found devouring sheet music at every parlor piano in the land (150). One wonders what he would have made of *Guitar Hero* and *Rock Band*. These games have restored the pleasures of "musica practica" as mainstream popular practice, but they also complicate Barthes's hard distinction between listening and playing. They link the physical gestures of live musical performance with the reproduction of recorded songs.

I refer to this phenomenon as "schizophonic performance," building on R. Murray Schafer's term for the split between a recorded sound and its source:

"This dissociation I call schizophonia, and if I use a word close in sound to schizophrenia it is because I want very much to suggest to you the same sense of aberration and drama that this word evokes, for the developments of which we are speaking have had profound effects on our lives" (1969:43). Schizophonia is not a value-neutral term, and some have questioned its suitability for describing sound-reproduction technology. Jonathan Sterne notes that the schizophonia concept is predicated on the notion that "face-to-face communication and bodily presence are the yardsticks by which to measure all communicative activity," so sound reproduction is "doomed to denigration as inauthentic, disorienting, and possibly even dangerous by virtue of its 'decontextualizing' sound from its 'proper' interpersonal context" (2003:20–21). I agree with such critiques, and I use the term "schizophonic performance" precisely because it captures the ambivalence and even paranoia that characterizes much *Guitar Hero* and *Rock Band* reception. In the twentieth century, musicians and audiences gradually came to terms with sound-reproduction technology by establishing value-laden distinctions between the live and the recorded (Wurtzler 1992, Auslander 1999, Waksman 1999). These games threaten the sanctity of those distinctions. They dramatize schizophonia, endorsing the idea of a split between the live and the recorded and inviting people to play at mending that split. To many critics, players' efforts are prefigured as "inauthentic, disorienting, and possibly even dangerous."

The title of this chapter invokes a classic text in ethnomusicology: John Blacking's 1973 book *How Musical Is Man?* Much of that book was devoted to arguing that the musical traditions of the Venda people of South Africa and those of Western art music are equally valuable, equally culturally contingent, and equally human—that is, that musical expression is rooted in human cognitive structures and realized through the mobilization of the human body. In framing his argument, Blacking suggested that "the reassessment of human musicality" was ethnomusicology's most pressing task (1973:4). Almost 40 years later, *Guitar Hero* and *Rock Band* gameplay both invoke and apparently threaten some deep-rooted beliefs about authentic musicality, creativity, authorship, and performance. It might take a culture-wide reassessment of human musicality to vouchsafe wide agreement that what players do with music in these games should even count as "musical."

VIRTUAL ROCK AND ITS DISCONTENTS

The first *Guitar Hero* game was created by the game development company Harmonix Music Systems in response to a commission from RedOctane, a company that made dance pad peripherals for the *Dance Dance Revolution* video games.[1] RedOctane wanted to create a market for a guitar-shaped game controller, inspired in part by a similar controller for Konami's *Guitar Freaks* (a Japanese arcade game that has not been released in the United States).[2] *Guitar Hero* (2005) and *Guitar Hero II* (2006) were extremely successful, attracting the attention of multinational corporate suitors. In 2006, Harmonix was acquired by MTV Games (a Viacom subsidiary) and RedOctane was acquired by Activision (a Vivendi subsidiary), creating a split between *Guitar Hero*'s developer and its publisher. Activision turned the *Guitar Hero* franchise over to another development company, Neversoft, which produced *Guitar Hero III* and subsequent editions—putting the Harmonix team in the curious position of designing their next game, *Rock Band*, as a head-to-head competitor with the series they had originally developed. By the end of 2010, the *Guitar Hero* and *Rock Band* franchises included 10 major releases. (See Table 3.1.)

The early *Guitar Hero* games included guitar and bass parts; *Rock Band* and later editions of *Guitar Hero* added a drum kit and karaoke vocals (see Figures 3.1 and 3.2).[3] A single player may choose to play any one of these four musical lines, or up to four people may play together, forming the classic four-piece rock band lineup. (Some players use a mic stand so that one person can play an instrument and sing karaoke simultaneously.) In each game, players choose their performance repertoire from a long list of popular songs—mostly rock, punk, and metal, dating from the 1960s to the present (see Table 3.2). The guitar, bass, drum, and vocal parts of each song have been transcribed at four difficulty levels using a special notation system. Additional

Table 3.1 *GUITAR HERO* AND *ROCK BAND* GAME EDITIONS

game	developer	publisher
Guitar Hero (2005)	Harmonix	RedOctane
Guitar Hero II (2006)		
In 2006, MTV Games bought Harmonix; Activision bought RedOctane.		
Activision turned Guitar Hero development over to Neversoft.		
Guitar Hero III (2007)	Neversoft	Activision
Guitar Hero World Tour (2008)		
Guitar Hero 5 (2009)		
Guitar Hero: Warriors of Rock (2010)		
Rock Band (2007)	Harmonix	MTV Games
Rock Band 2 (2008)		
The Beatles: Rock Band (2009)		
Rock Band 3 (2010)		

Additional series expansion releases include *Guitar Hero Encore: Rocks the 80s* (2007), *Guitar Hero: Aerosmith* (2008), *Guitar Hero: Metallica* (2009), *Guitar Hero Smash Hits* (2009), *Band Hero* (2009), *Lego Rock Band* (2009), *Green Day: Rock Band* (2010), and several hand-held/mobile phone releases.

Table 3.2 ROCK BAND REPERTOIRE

genre	# of songs	% of catalog
Rock Band (2007)		
Alternative	9	15.5%
Classic Rock	6	10.3%
Emo	1	1.7%
Glam	2	3.4%
Metal	4	6.9%
Pop-Rock	3	5.2%
Progressive	2	3.4%
Punk	2	3.4%
Rock	26	44.8%
Southern Rock	3	5.2%
Total songs	58	
Rock Band 2 (2008)		
Alternative	11	13.1%
Classic Rock	10	11.9%
Grunge	5	6%
Indie Rock	4	4.8%
Metal	12	14.3%
Nu Metal	3	3.6%
Pop-Rock	12	14.3%
Progressive	4	4.8%
Punk	6	7.1%
Rock	13	15.5%
Southern Rock	1	1.2%
New Wave	3	3.6%
Total songs	84	
Rock Band and Rock Band 2 songs, by decade		
1960s	3	2.1%
1970s	30	21.1%
1980s	21	14.8%
1990s	32	22.5%
2000s	56	39.4%
Total songs	142	
Rock Band total catalog as of November 2010 (all game editions and DLC)		
Alternative	266	12.4%
Blues	15	0.7%
Classic Rock	186	8.7%
Country	54	2.5%
Emo	31	1.5%
Glam	14	0.7%

Table 3.2 (continued)

genre	# of songs	% of catalog
Grunge	57	2.7%
Indie Rock	119	5.6%
Jazz	2	0.1%
Metal	330	15.4%
Novelty	13	0.6%
Nu Metal	39	1.8%
Other	65	3%
Pop-Rock	205	9.6%
Progressive	48	2.2%
Punk	137	6.4%
R&B Soul/Funk	6	0.3%
Rock	466	21.8%
Southern Rock	26	1.2%
Urban	15	0.7%
New Wave	42	2%
Total songs	2137	

This data is compiled from the song catalog published at RockBand.com. Note that the *Rock Band* repertoire was intended to reach a broader demographic than the *Guitar Hero* repertoire; it also had to be relatively singer-friendly, due to the addition of the vocal part. The first three *Guitar Hero* games included very high proportions of rock and metal songs.

songs or entire albums can be downloaded for a fee, and new downloadable content is released regularly. By 2010, total sales revenue for the two franchises had exceeded $3 billion. This figure represents sales of over 46 million game units and over 115 million downloads of additional songs for the games, echoing the popularity of mass-produced sheet music in an earlier era.[4] Adults and children alike are avid players; a 2010 media study found that 71 percent of American 8- to 18-year-olds had played *Guitar Hero* or *Rock Band* (Roberts, Foehr et al. 2010).

To play a song, a *Guitar Hero* guitarist must read a moving stream of on-screen notation, simultaneously pressing a particular fret button with the left hand and the strum bar with the right hand as each note crosses the bottom of the screen (see Figures 3.3 and 3.4). The player functions as the gatekeeper for prerecorded material; correct fretting/strumming allows each note to make its way from the game console to the speakers. If the player misses a note, that note drops out of the audio playback. (At the lower difficulty levels, a single on-screen note may correspond to a short riff on the recording.) At the borders of the streaming notation track, animated avatars perform the song in a rock club or arena. Frenetic camera angles and stage effects add to the excitement. The virtual crowd shouts praise to good players, boos incompetent players off the stage, and sometimes sings along.

Figure 3.1
Rock Band guitar controller. (*Photograph by the author*)

Figure 3.2
Rock Band drum controller. (*Photograph by the author*)

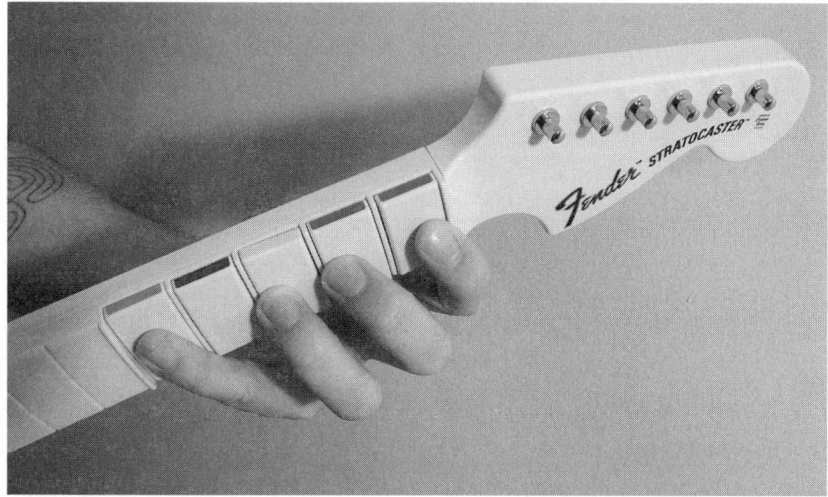

Figure 3.3
Color-coded fret buttons on the guitar controller neck. *Rock Band* guitar controllers include a second set of fret buttons close to the body of the guitar for extended play techniques ("tapping") and to make fretting easier for small children. Note that the player must change hand position or extend a finger to access all five buttons. (*Photograph by the author*)

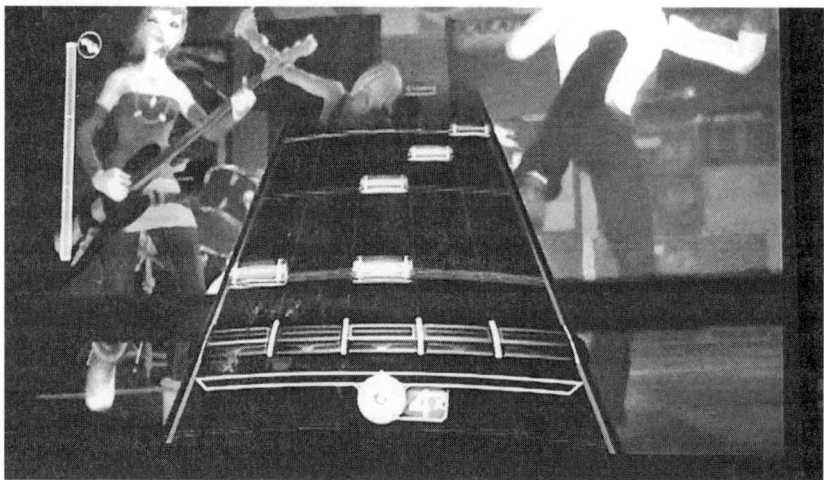

Figure 3.4
Rock Band notation. (*Screenshot by the author*)

Figure 3.5 is a schematic diagram of *Guitar Hero* and *Rock Band* guitar/bass notation. Imagine the notes in the diagram falling from the top of the page to the bottom, but with perspective applied so that they appear to be coming directly at you, like the road in a driving game. The letters in the diagram indicate the color of the note (green, red, yellow, blue, orange). These colored notes

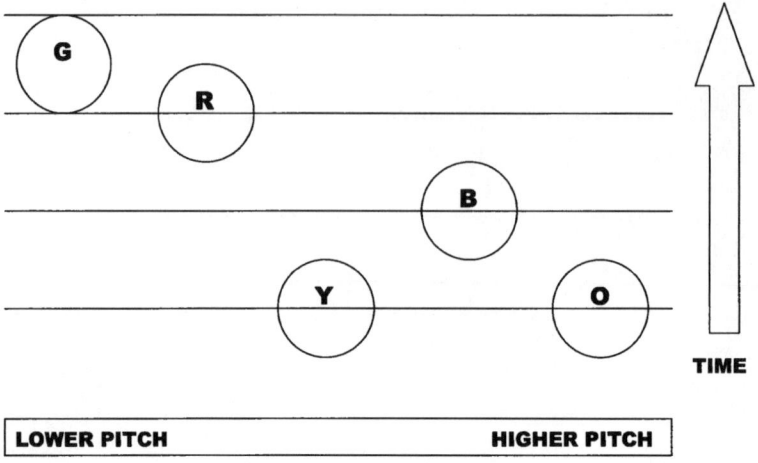

Figure 3.5
Schematic diagram of *Guitar Hero* and *Rock Band* guitar/bass notation. In the games, sustained notes are represented with a long tail following a note head. Smaller note heads designate "hammer-on" notes, which may be played by fretting without strumming. (*Diagram created by the author*)

mirror the layout of the five colored fret buttons on the guitar controller. As each note—or several notes, in the case of power chords—crosses a fixed reference line of colored notes at the bottom of the screen, the player must fret and strum. (In drum notation, there are only four note columns, corresponding to four color-coded drumheads. A horizontal orange bar across the staff indicates the kick drum.)

This notation system reverses the conventions of Western staff notation. Here, the horizontal axis represents relative pitch and the vertical axis represents the passage of time. The horizontal lines correspond to strong beats; notes between the lines subdivide beats. There is no uniquely correct staff-notation transcription of the diagram in Figure 3.5, because melodic contour and rhythmic relationships are represented only in relative terms. However, any competent player would read the diagram from bottom to top and would know that it represents a two-note chord (YELLOW + ORANGE) followed by a descending melodic line of three notes (BLUE, RED, GREEN), most likely with a melodic skip from BLUE to RED and stepwise motion from RED to GREEN. The player would also surmise that this notation is from a song being played at the "hard" or "expert" difficulty level, because it employs all five fret buttons. The "easy" version of a guitar part uses only green, red, and yellow; "medium" adds blue; "hard" and "expert" add orange. Because the player has only four fingers available for fretting, she or he must change hand positions when the notation requires all five fret buttons—a design decision that creates technical fingering puzzles for players to solve. As Dominic Arsenault notes in an article assessing

Guitar Hero as a guitar simulator, there is "an emphasis on breadth and a deliberately limited depth" in the simulation; the designers' apparent strategy was "to represent a little of everything" (2008; cf. Blaine and Fels 2003:129).

These games have some important schizophonic forebears, including lip-syncing, air guitar, and karaoke, but each of these performance genres has distinct features. Karaoke singers use prerecorded accompaniments, but they are still singing with their own voices. *Guitar Hero* and *Rock Band* players are instead serving as gatekeepers for *someone else*'s musical performance (and they're often sight-reading songs they've never heard or closely listened to before, which is virtually never the case with karaoke). Lip-syncing or air guitar might seem like a better match, but that comparison completely ignores gameplay mechanics. The reward system for these games relies on mastery of a specific technical interface: If players stop playing their instrument controllers or play them inaccurately, they will disrupt or destroy the game's musical output. *Atlantic* writer James Parker describes this experience vividly: "I hit [the note] once, I hit it twice, and then . . . I miss it. Damn! The [on-screen] notes keep coming, but my '70s supersound has broken up—what? how?—into clucks and clangs of dissonance, ruinous postpunk fragments. The song lurches and dies. Drones of disapproval from the crowd. . . . Numb with culpability, I stare at my Les Paul" (2009:36).

Parker's account indicates how these games inculcate a sense of responsibility for the musical performance, thereby bridging virtual and visceral performance (cf. Grodal 2003). As one player told me, "You are delivering that performance to people. . . . You're performing a song that's already been made before, but without your input, it wouldn't sound as good. And so, you're partially responsible for that musical experience for people."[5] Some players seem to think of themselves as spirit mediums, channeling an ancestral musical performance. One person wrote in a survey response, "I feel like I'm jumping into the artist in their time and playing along and maybe even feeling what it was to be that creative individual in their time" (iRone, male, 31–35).[6] An interviewee observed, "These are songs that your heroes or your music idols might've played on stage and you might've seen them. . . . I'm sure there's some sort of association between those feelings you get watching someone play a song on the stage live, and the feeling you get playing it through Rock Band."[7]

To critics, all this is simply evidence that the games perpetuate escapist delusions. Channeling Adorno instead of Jimi Hendrix, they depict players as dupes of the mass-culture industry, feverishly pressing buttons in order to indulge in a fantasy of rock stardom that relies on the mechanized reproduction of commodified hits. As Parker wrote in *The Atlantic*, "Games are games, but you don't have to be Jean Baudrillard to detect in all this another coup by the forces of unreality—a decisive one, perhaps" (2009:37). Numerous journalists, professional musicians, and online commenters have expressed disdain, discomfort, or wistful regret about the whole *Guitar Hero* phenomenon. Commenters on thousands of *Guitar Hero* and *Rock Band* YouTube videos evince a deep concern with the authenticity of players' performances, not only by exhorting players to repeat their feats on real instruments but by accusing them of faking high game scores through technical trickery. Even fans of the games often express mixed feelings about high-level players. As one player told me,

> Even though when you are playing you wish, "Oh, I wish I was as good as those guys," but when you watch them do it, it's kinda like, "How much time did you spend doing that?" . . . Those guys—you can also look at their appearance and see, like, the Cheeto stains on their shirts. . . . And the line of dialogue associated with that is "Dude, just learn how to play the instrument." . . . I mean, I guess they are tricking themselves into believing, like, "Oh, I can five-star this song," and it's much easier than to actually learn to play the guitar.[8]

Faced with this distressingly ambiguous musical practice, many people ask players, "Why don't you pick up a real guitar?" This is such a common reaction to the games that online commenters often refer to it with the shorthand abbreviation "lrn2reeltar," as in this irony-laced messageboard exchange:

> Gabrie|a: Maybe some people play GH and play the guitar:O OMG?!

> reply: lies!

> second reply: Actually I do both . . .
> And to all you people saying to lrn2reeltar.
> Why don't you go take all the CoD players and tell them to join the army.
> Why don't you tell all the Madden players to go play in the real NFL.
> (MySpace Forums 2008; the third writer is referring to the war-themed *Call of Duty* series and the football-themed *Madden NFL* series.)

Celebrity musicians like Jack White, John Mayer, and Slash periodically weigh in, fueling renewed media coverage of the games and new flurries of online debate. When Jack White told reporters, "It's depressing to have a label come and tell you that [*Guitar Hero*] is how kids are learning about music and experiencing music," online commenters used his remarks as a springboard for airing their own views:

> Bigcitythekidd: i believe what they are trying to say is, that kids should be going out and playing REAL instruments. Kids now a days want to be good at guitar hero rather than suffer and put time into a real instrument. the feeling when playing music and going to that place .. "that place" is far greater than any feeling from a video game.

> Prariwolf: I too thought Guitar Hero was a big waste of time when my nephews got it for Christmas a couple of years ago. Instead, it got them interested in

these guitar greats, looking up and downloading their stuff from iTunes and other sites (all legal, no worries). Then my older nephew asked for a guitar this year, so he could REALLY learn to play. So now I see it as a stepping stone. If there is no real interest, it's a harmless game. If there is, it's a good place for a beginner to get exposed to the art.

(comments posted June 23, 2009, on Luerssen 2009)

A related line of critical discourse uses *Guitar Hero* to shore up the ideological distinction between rock and pop, often along gender lines. For instance, one website juxtaposed an iconic photograph of Johnny Cash making an obscene gesture at the camera with a photograph of Miley Cyrus smashing a *Guitar Hero* controller (in the manner of Pete Townshend or Jimi Hendrix). The caption read: "Everything that's wrong with pop culture in two photos." One commenter wrote, "The pussification of the world continues." Another surmised, "Maybe she (Miley) feels threatened by guitar hero, for it is part of the few things that approximate her cultural eradication potential" (John 2010).

As these examples suggest, much of *Guitar Hero* and *Rock Band* reception has revolved around matters of authenticity, including assessments of authentic musicality, authentic skill, authentic effort, and authentic masculinity. In this chapter I address some of these reception discourses and compare them to players' accounts of their experiences with these games. How do game-design decisions, media coverage, and preexisting discourses about rock authenticity, liveness, and gendered performance influence those assessments? How is experiencing a prerecorded rock song through virtual performance different from simply listening to that song? Why *don't* all *Guitar Hero* players simply pick up a real guitar? It's easy to imagine why a *Call of Duty* player might not join the army; perhaps she or he is politically opposed to war, doesn't want to risk getting shot, or simply has other career goals. Similarly, only a few of the millions of *Madden NFL* players have the athletic potential and training resources required to become professional football players. But most people with the financial and physical means to acquire and play *Guitar Hero* or *Rock Band* could easily choose to take up instruments instead. Why don't they? Moreover, my research has shown that many players do both. What can those players tell us about the appeal of these games, and about how playing a game controller compares to playing a musical instrument?

I have investigated these issues by interviewing game designers and players, analyzing responses to an online qualitative survey, reading player message-board discussions, attending tournaments, watching YouTube videos, trying to keep up with a flood of print and online media coverage, and playing the games myself in a variety of settings (e.g., at home alone, with my spouse, in party situations, and at work). For six weeks in the summer of 2008 I developed my *Rock Band* guitar and drum skills through daily practice, both on my own and with Kate Reutershan, my undergraduate research assistant. We recruited 15 Brown University undergraduates and graduate students for gameplay/interview sessions (advertised by posting flyers around campus). These students were all experienced *Guitar Hero* or *Rock Band* players. While we made no effort to recruit players with formal music training, 14 of the 15 also had significant experience playing musical instruments. The volunteers were mostly math, science, and engineering students with summer lab jobs; only one was a music major. They were 13 men and 2 women, of diverse ethnic and musical backgrounds: for instance, a Mexican American electric guitarist, two white American rock drummers, an African American jazz saxophonist, a South Asian American bass player, a white American classical pianist, a classical pianist from Hong Kong, and a white American acoustic guitarist/ singer-songwriter. In the summer of 2009, I learned more about social, public, noncompetitive gameplay by attending *Rock Band* nights at various bars around Boston and interviewing organizers and participants. I also attended a few game tournaments and watched televised competitions during this period (see Chapter Four). At the bar nights, the gender split of participants was close to even; at tournaments, competitors were virtually all male.

My online survey about *Guitar Hero* and *Rock Band* was active from 2007 to 2010 (Miller 2007b). I received 547 responses and conducted follow-up e-mail correspondence with a 5 percent sample of these respondents. (This sample matched the total pool with respect to gender, age ranges, and proportion of yes/no answers to some key questions.) My survey and interview questions focused on *Guitar Hero* and *Rock Band* gameplay contexts, the games' impact on players' musical tastes, and comparisons to other musical experiences (e.g., playing real instruments or singing karaoke). Survey respondents were self-selecting; most heard about the survey through recruitment messages posted on several high-traffic online messageboards devoted to these games, or received the link from a friend who had already completed the survey. There is no reason to assume that these players represent an ideal demographic sample of the millions of people who have purchased or played the games.[9] However, the range of experiences represented in the responses to my qualitative questions does seem to match the range of perspectives I encountered through other research channels. A few statistics should give a sense of the survey respondent pool and the key questions:

- 85 percent were male; 14 percent were female; 1 percent were intergender/trans/other.
- 56 percent were age 21 or younger; 26 percent were age 22 to 30; 19 percent were over 30.[10]
- 99 percent had played some version of *Guitar Hero;* 48 percent had also played some version of *Rock Band;* 1 percent had played only *Rock Band.*[11]
- 90 percent owned some version of *Guitar Hero;* 29 percent owned *Guitar Hero* and *Rock Band;* 4 percent owned only *Rock Band.*

- 45 percent typically played for 1–2 hours at a time.
- 18 percent usually played at the "easy" or "medium" difficulty level; 21 percent usually played at "hard"; 61 percent usually played at "expert."
- 71 percent had used "practice mode" (which breaks songs down into short sections that can be drilled at slower tempos).
- 53 percent often played with other people watching; 69 percent often played in a multiplayer mode.
- 59 percent reported that they knew fewer than half of the songs in the games very well before encountering them during gameplay.
- 78 percent had added new music to their listening collections because of the games.
- 74 percent had experience playing an instrument.
- 48 percent had experience playing guitar; 33 percent had played in a band; 14 percent regularly performed music in public.
- 36 percent indicated that they felt creative during gameplay.

Those who assert that these games offer only hollow simulacra of musical performance might be surprised to learn that three-quarters of the survey respondents and 14 of the 15 gameplay/interview volunteers had experience playing an instrument. The larger lesson of these statistics is that America remains full of "hidden musicians" (Finnegan 1989). Musica practica hasn't died out after all; to the contrary, a national telephone survey conducted in 2009 found that 58 percent of American households were home to at least one person who played a musical instrument (NAMM 2009:132). Still, it seems notable that the gamers who participated in my research had even *more* experience playing instruments than the general population.

This fact stands in intriguing tension with the mission statement often repeated in media interviews with designers at Harmonix Music Systems, the company that developed *Guitar Hero* and *Rock Band*: "to give that awesome feeling [of performing music] to people who aren't musicians, who would never get to have it" (here articulated by audio director Eric Brosius; Dahlen 2008). Such statements are meant to broaden the reach of the games. Harmonix designers are certainly aware that many musicians love the games, and quite a few Harmonix staff members play in rock bands themselves. As we shall see, however, critics of the games have often assumed that there is a fundamental split between people who play controllers and people who play instruments.

"WON'T GET FOOLED AGAIN": EVALUATING AUTHENTIC MUSICALITY

The word "real" comes up over and over in press coverage of these games and in responses to my survey. For instance, a *San Francisco Chronicle* article is titled "Rock Band, Guitar Hero III Video Game Do Rock, but Real Is Better"

(Hartlaub 2007); an article in *Guitar Player* magazine about the cover musicians who recorded some of the in-game songs is titled "The Real Heroes of *Guitar Hero III*" (Ross 2008); and a review of *Rock Band* published in *Spin* appears under the heading "Even Better than the Real Thing" (Anderson 2007). These value judgments bring to mind the lip-syncing scandals that have beset popular music performers from time to time. As Philip Auslander writes, lip-syncing threatens rock ideology by suggesting the possibility of "a new era of music performance in which the visual evidence of performance would have no relation to the production of sound" (1999:86). *Guitar Hero* and *Rock Band* seem to represent the realization of this threat. Performers go through the motions of playing instruments, but they have no creative dominion over the song that comes out; it was originally produced by some other guitarist in a recording studio.

One might identify precursors to this phenomenon in the world of electronic art music. By the mid-twentieth century, composers were experimenting with what Leta Miller calls "a model of cooperative interdependency" among composers, musicians, dancers, and audio engineers. In works such as John Cage's and Merce Cunningham's *Variations V* (1965), "the interaction of sound and motion was facilitated by a sophisticated technological component" (Miller 2001:547). In such works, performance processes were sometimes defined with great specificity, but in most cases "the resulting sound remained variable" (546). In *Guitar Hero* and *Rock Band,* by contrast, very different physical performances may result in the same sound output—a fact that brings to mind Jacques Attali's argument that "representation has become an auxiliary of repetition" in our present era (1985 [1977]:85). As Henry Adam Svec observes, "The player can either conform to the game's logic by reproducing the requisite hits, which are presented as measurable, stable, complete, and eternal (*structural*), or not play at all" (2008).

The more a player's physical performance matches that of a live performing musician, the more unsettled some audience members become. Their distress brings to mind an Adornian nightmare, in which "popular music divests the listener of his spontaneity and promotes conditioned reflexes," and "pseudo-individualization" hides standardization, "endowing cultural mass production with the halo of free choice" (Adorno 1941:22, 25). Adorno put forward the idea that in popular music "the composition hears for the listener," by establishing repetitive patterns that encourage simple recognition rather than serious musical listening (22). *Guitar Hero* seems to go a step further, making it possible for the composition to play for the performer.

Media critics and professional rock musicians have advanced similar arguments about the games. A *San Francisco Chronicle* writer suggested that "something . . . seems fundamentally wrong when you pick up the video games. . . . What kid will ever want to pick up a real guitar, when learning to play a fake one is so easy? If Rock Band had been available in the late 1980s, would we even have a Green Day—or just three more no-name slackers killing a lot of time in their parents' basement?" (Hartlaub 2007). Sleater-Kinney guitarist Carrie Brownstein asked in a *Slate* article, "Really, if you are going to play the game with a group of friends for more than a night, shouldn't you just form a real band?" (2007). Guitarist John Mayer was more pointed when a *Rolling Stone* interviewer asked what he thought about these games: "Guitar Hero was devised to bring the guitar-playing experience to the masses without them having to put anything into it" (Hiatt 2008). An online commenter on this interview concurred: "'Guitar Hero' and 'Rock Band' are the epitome of what's wrong with twenty-first-century pop culture. It's all easy, hollow, and accessible to any fucking moron" (jcurrier1981, comment posted June 4, 2008).

In *Pop Music and the Press,* Steve Jones and Kevin Featherly interrogate the concepts of authenticity most often employed in rock criticism. As they note, "Since standards are difficult to come by in popular music, critics often refer to authenticity as a measure of aesthetic soundness" (2002:32). Their survey of forms of authenticity includes the authentic articulation of the ideas or desires of a particular audience (i.e., not selling out/pandering to the "mainstream"); historical authenticity (roots-oriented music); cultural or ethnic authenticity (being true to one's musical culture of origin); and personal authenticity rooted in self-expression (32–33). As many scholars and critics have demonstrated, this last standard generally relies on assessments of both authenticity of intention and authenticity of creative artistry. Personal authenticity derives from "a perceived quality of sincerity and commitment" (Frith 1996:71), "some kind of unmediated openness of expression" (Christgau 2004:37), and heroic genius, "the myth that had first arisen in the 1820s of the creative individual, uncompromising in passion and bearing unique gifts" (Weinstein 2004:191). Nearly all the repertoire included in *Guitar Hero* and *Rock Band* has received critical endorsement in accordance with one or more of these standards, either on the music's original release or in the course of its reception history; indeed, this "authenticated" status is what qualified songs for inclusion in the games. To many critics, though, this fact only enhances the offensiveness of *Guitar Hero* and *Rock Band* gameplay. Musical authenticity is not just "a vague sensibility or aesthetic" but "a cultural value anchored in concrete, historical practices of production and consumption" (Thornton 1996:5). Among rock musicians in particular, "technologies and technological adaptations are tied to cultural practices that authenticate musicianship and signal alliances within and alienation among musical and social groups" (Gay 1998:81). A mass-market video game that simulates the performance of authentic rock music potentially undoes all that authenticating work, devaluing the repertoire's subcultural capital. Musical works that formerly represented creative genius, technical mastery, and sincere commitment become "easy, hollow, and accessible."

Players and game designers hear and respond to these arguments all the time, and many pass similar judgments: They want to make it clear that they understand the difference between authentic musicianship and video game prowess. Defenders of the games typically rely on four key points. First, they assert that *Guitar Hero* and *Rock Band* are just games and that no one is trying to say they represent real musicianship. For example:

> It isn't supposed to be a real instrument and shouldn't be compared to one. It is for fun, like every other game. Why play racing games when you can go out and drive your car? Why play Guitar Hero when you can just learn a real instrument? Why play RPGs [role-playing games] when you can go stab squirrels in your back yard? Why do we even have video games if everyone can just do the real thing? (Stinky89, comment posted July 7, 2008, on Stallock 2008)

Greg LoPiccolo, vice president of product development for Harmonix Music Systems, made a similar distinction:

> You know, a lot of people have very visceral reactions to these games. They don't like them. They think, "Well, why not play a real guitar?" And I think the answer to that is, well, you *should* go play a real guitar. I mean, the people at Harmonix are happy, the more people take up real instruments the happier we are. But it's not—you're not performing music, you're playing a game. It's a different experience. (DeRogatis and Kot 2009)

Second, defenders argue that the games offer an engaging, interactive musical experience to people who are "not musical" and would otherwise never get the chance to experience the joys of musical performance. Heather, a woman in her 20s who often played *Rock Band* at bars in Boston (see Chapter Four), was among the players who identify themselves as "unmusical"—though her own account of her gameplay experience contradicted that assessment:

HEATHER: I started playing Rock Band, and I was like, I love this. It is amazing. . . . Rock Band, having all the instruments together, makes me play better [as opposed to just playing the guitar part in *Guitar Hero*], makes me understand the beat of it better. . . . I listen to the other instruments, and it helps me formulate the song in my head for the next notes for the bass or the guitar coming along. . . .

KM: So do you have any previous musical training on instruments or anything like that?

HEATHER: No, music and me does not go. . . . There's people out there that are gifted with music, and I'm not one of them. . . . The people that are real musicians wouldn't want to play [*Rock Band*] because it's too easy and infantile.[12]
▶ 3.1

Heather has apparently adopted "a generalized role of musical outsider" (Kingsbury 1988:74), although she loves music and derives great pleasure from playing *Rock Band*. She has identified herself as the target consumer for this game: it might be satisfying for an "unmusical" person, while holding little appeal for people with inborn musical talent.

Third, defenders suggest that the games could offer a stepping-stone to "really playing music," because they develop players' sense of rhythm and might inspire them to pick up real instruments. On the Harmonix website, company CEO Alex Rigopulos bridges the "unmusical" and "stepping-stone" arguments, assuring potential customers that the games are accessible to everyone while also emphasizing that he still believes in the primacy of "old-fashioned" musicianship:

> Everyone is born with an innate urge to make music. It's one of the most profoundly joyful things in life. Yet the challenges are such that few people really get that far. We thought that was a significant problem, and we set about exploring new ways to solve it. Our mission was to show nonmusicians how it feels when you finally get to the other side. And hopefully, to inspire them to start making music the old-fashioned way. (Harmonix Music Systems 2010a)

This is an argument that appeals particularly to adults who might purchase these games for children. Like "Prariwolf," who testified to *Guitar Hero*'s beneficial musical influence on his/her nephews, many adults are happy to see children engaged with a video game that seems at worst harmless and at best educational, "a good place for a beginner to get exposed to the art" (comment posted on Luerssen 2009; see full quote earlier). Travis Stever, a guitarist whose songs are featured in the games, compared *Rock Band* to *Grand Theft Auto*, remarking to a reporter, "Better [that kids] start bands than kill people and beat up prostitutes, right?" (Anderson 2007:110).

Finally, proponents often point out that the games make players better musical consumers, because they help people improve their listening skills, develop new respect for the musicians who recorded the songs in the games, and discover new songs, artists, and genres. Mike Dadmun was such an engaged player at *Rock Band* bar events in Boston that the event organizer recruited him to run events of his own. He told me,

> What it's actually allowed me to do is get involved and discover new bands. I've downloaded so much music, just from Rock Band itself, and hearing the different music that's in there, and I've fallen in love with new bands. . . . I think you become more inclined to relate to the artist and to the music because you're actually involved in the game and the music itself. So if you just listen on the

radio, you might just change the channel when you hear [an unfamiliar song or genre]. But I think when you're playing with other people, and other people want to play a song you don't know, and they're playing with the hardware and listening to music, they actually end up liking that song and liking that artist they've never heard.[13]

So to sum up: these games are just games, they shouldn't be compared to really playing music, and people shouldn't take them too seriously—but they also have all sorts of potentially positive effects on players' musical sensibilities and future musical pursuits. It's a complex and often contradictory argument. In their efforts to defend themselves against charges of delusions of musicianship, a lot of players seem intent on *not* reassessing what counts as authentic musicality. Devon O'Dell, a competitive tournament player who referred to *Guitar Hero* as "my major activity outside of my job," is among those who defend the games by disavowing their musicality:

> I love music, and I love making music (however good or bad I may be at that notwithstanding). . . . Comparing a game, which one may or may not take very seriously, with a [musical] hobby or profession . . . is, quite frankly, comparing apples to oranges. . . . Nobody who plays the game—whether they're serious or simply playing for fun—is under the impression that playing the game is anything like playing a "real" instrument. . . . Guitar Hero and Rock Band are nothing more than hand–eye coordination games that reward your ability to push buttons rhythmically. People may or may not find bliss in doing that, but actively despising an entire franchise and millions of people who play the game because they are doing something they enjoy seems counterintuitive to me. (survey follow-up email correspondence)

These apples-and-oranges, chalk-and-cheese, "just a game" defenses are common among players who would prefer to lift the games straight out of the ideological morass of "real music" comparisons. Their points are well taken; the quality of digital gameplay should be evaluated based on a game's design merits and its players' experiences, not through comparisons that treat it as an imitation of some other activity. However, my research has shown that the boundary between playing these games and "really playing music" can be quite porous. (Indeed, O'Dell also told me that he had purchased a Roland electronic drum kit, modified it to serve as the drum controller for *Rock Band*, and used it to compose percussion parts on his computer; he was looking into taking private drum lessons.) The ways players approach these games and talk about their experience indicate that *Guitar Hero* and *Rock Band* engender undeniably real emotional, analytical, social, and physical engagement with music.

Players often struggle with the apparent contradiction between *feeling* like they're really playing music, even playing creatively, and acknowledging that they're functioning as middlemen for a prerecorded track. Steffen, a math graduate student with years of experience as a rock drummer, contrasted the feeling of playing music in these games with that of simulated warfare in other games:

> So there's a big difference between pressing X and having someone shoot someone else on the screen, and pressing X a couple times and successfully putting out a guitar riff. Even though you haven't actually put out the guitar riff, the game makes you feel like you have, and that's a really weird sort of feeling of creativity that maybe is not genuine. Maybe, in some sense, it's not real creativity. It's you doing what the game wants you to, but the feeling you get inside. . . . It's certainly an accomplished feeling. It makes you feel like you actually played the song.[14]
>
> ▶ 3.2

Steffen's remark about the "big difference" between first-person shooter games and *Rock Band* seemed cryptic to me at first, but it highlights a crucial experiential distinction. Shooting another character in a video game doesn't put a real bleeding corpse in your living room, but pressing buttons or hitting drum pads on game-controller instruments brings forth real music. This "realness" is grounded in a shared cultural assumption that music consists first and foremost of sound, so recorded music is not considered to be "fake" music (although live musical *performance* retains special status, and faked live performance therefore evokes special scorn).

The "realness" of the music in *Guitar Hero* and *Rock Band* is constituted in two ways: first, through players' respect for the aesthetic quality, technical difficulty, and emotional power of the original recordings coming out of the speakers; second, through the games' capacity to inspire in many players the *feeling* of making music. For instance, when I asked a player who had previously dabbled in trumpet, saxophone and guitar, "To what extent does *Guitar Hero* feel like making music to you?," he replied:

> It feels pretty good. It feels like I am doing something even though I know, like, logically speaking, it's just a computer program waiting for me to hit the button. It's the simplest of all games, really, hit the button, hit the button, hit the button. . . . But from a very instinctive point of view, like as a human, it feels right, and it feels like I am doing something worthwhile. Even though it is just a video game, it just feels right.[15]
>
> ▶ 3.3

What accounts for this compelling, satisfying feeling, which many other players characterized as "getting lost in the music"? Imaginative, immersive role-play of a rock-star fantasy is certainly part of the answer for some players, and it is well supported by the games' audio and graphic design. As Harmonix audio director Eric Brosius put it, "Everything points to the same goal, which is that you're not sitting in your living room holding a plastic toy, you're a rock star on stage" (Dahlen 2008). However, only a handful of my survey respondents and interviewees suggested that this was the most important factor in their experience of the games. Instead, they talked about the nuts and bolts of *Guitar Hero* and *Rock Band* musicianship: reading streaming notation, translating it into instrumental technique, and listening to the song unfold.

VIRTUAL AND VISCERAL MUSICIANSHIP

Descriptions of *Guitar Hero* gameplay often reduce the basic game activity to "pushing buttons in time." This summation is not inaccurate, strictly speaking, but it fails to capture the feeling and the appeal of gameplay for most players— much as it would fail to capture the feeling of playing a Chopin nocturne at the piano. If one sets aside all of the rock-related framing devices—the instrument controllers, the on-screen avatars, the repertoire—the core challenge-and-reward system in these games is a sight-reading simulator. (Indeed, many YouTube videos of gameplay include "sightread" in the title, as a disclaimer or a boast.) The notation streams toward you at a constant pace. If you miss a note, you don't hear that note. Depending on the version of the game, you might hear a "mistake" indicator instead—the clank of a mangled guitar note or the dull tap of a misplaced drum hit. The rest of the band plays on; only your line drops out. This design decision plays a huge role in the overall gameplay experience, because it provides sensory evidence that the player is producing the sounds that come from the speakers. When I asked one player whether playing the games ever felt like playing music to him, he replied, "Well, when it gives you those screeching bad notes when you miss, then yeah, it's like 'Oh, I better do this right so I can make the music.'"[16]

Each song in the games can be played at one of four difficulty levels: easy, medium, hard, or expert. Attempting a guitar, bass, or drum part on the "easy" setting, players see and play notation that corresponds to a skeleton of the music that comes out of the speakers; playing on "expert" entails note-to-note correspondence. As a player explained, "When you start on easy, basically you're hitting every third note that you're hearing. So medium would be like maybe every other note, and expert is pretty right on to what you're listening to. So as soon as people start playing, right off the bat, they start to want to hit a lot more notes than what they're doing."[17] This mismatch between "easy" or "medium" notation and the complexity of the part on the audio track creates an incentive for players to improve their technical skills and move up to the higher difficulty levels. Also, as Peter Shultz notes, the "reductive analysis" that the transcribers use to create the four difficulty levels "encourag[es] the player to hear and think about each level in terms of the ones they have already played. . . . The game becomes an advocate for a hierarchical order of musical awareness: Players conceive of difficult patterns as elaborated versions of simpler ones, in a manner analogous to Schenkerian and other reductive or generative theories of music" (2008:187).

At even the "medium" or "hard" setting, the pace of the streaming notation can be overwhelming to players who keep their eyes at the bottom of the screen. Without exception, the players I interviewed said that they read ahead. They kept their eyes on the new notation streaming from the top of the screen while using the sound of the music, their sense of rhythm, and occasional downward glances to play the notes that were passing over the bottom line. Most players said that they read the descending notes in chunks, looking for patterns and familiar material (e.g., scalar passages, drum patterns, or riffs that repeat from song to song). Several reported having learned this technique either from sight-reading standard notation or from playing *Dance Dance Revolution,* which employs a similar notation system to indicate dance footwork (Konami Corporation 1998, Smith 2004, Demers 2006). Players can also use "practice mode" to slow down and repeat particular sections of each song, which allows them to work out tricky fingering patterns for the guitar or learn difficult rhythms through focused listening at a slower tempo. However, they must always maintain *some* steady tempo, as though they were always playing in an ensemble or practicing with a metronome.

The transcription and notation systems employed in these games, along with players' notation-reading practices, offer striking parallels to a much earlier amateur musical practice: playing four-hands piano transcriptions. As Thomas Christensen writes, in the nineteenth century these transcriptions presented a "particular translation" of the contemporary concert repertory, one in which "mass textures . . . were often radically compressed" (1999:256). The transcriptions "played a surprisingly destabilizing role by blurring any number of traditional musical polarities," especially those distinguishing symphonic/chamber repertoires, professional/amateur status, active/passive music acculturation, masculine/feminine associations, and public/private reception spheres. They also contributed to the process by which "a coalescing canon of musical 'masterworks' was constituted and experienced by many musicians of the nineteenth century" (256).

Four-hands piano transcriptions were hugely popular among amateur pianists, but were criticized for their perceived "deleterious effects on piano pedagogy" (Christensen 1999:258). Students were always clamoring for new four-hands repertoire, which led to the production of some shoddy transcriptions that distorted melodic lines, voicing, and harmony. (Players voice similar complaints about "unmusical" transcription in some of the rushed-to-market downloadable content for *Guitar Hero* and *Rock Band.*) Still, advocates cited the practice's potential to contribute to players' musical education by inculcating a deeper knowledge of the original works (260)—like the *Spin* writer who asserted that *"Rock Band* and its ilk can provide gamers with an intimacy with the music that only comes from playing the tunes" (Anderson 2007:110).

Some may object that four-hands pianists can really play the piano, whereas *Rock Band* players are only pretending. But as Christensen notes, contemporaries did not credit the pianists' work as "real" performances of the musical works in question; a transcription was considered "a facsimile, a copy that at best was comparable to an engraving of an oil painting" (1999:275). Even when transcriptions were musically sensitive, the "varied colors and sonorities" of the original works were "filtered through the monochromatic timbre of the piano strings" (256), a process that some considered an "emasculation of the symphonic body" (273). Four-hands pianists might be technically competent parlor musicians, but they could never "really" play a symphony—nor could they even really listen to one, unless they attended a concert performance. Still, the prominent Austrian music critic Eduard Hanslick called four-hands piano "the most intimate, the most convenient, and, within its limits, the most perfect kind of domestic music making . . . a wellspring of enjoyment and instruction" (262).

I will return to the four-hands parallel when discussing the impact of *Guitar Hero* and *Rock Band* on players' musical listening experiences. But first, it's worth noting that unlike four-hands pianists—who are doomed never to really play a symphony orchestra—most of my *Guitar Hero* and *Rock Band* interviewees and survey respondents did have some experience playing traditional rock instruments. Many of them spoke or wrote eloquently about the parallels and distinctions between gameplay and playing an instrument.

Dan, a math graduate student, brought a mike stand to our gameplay/interview session so that he could sing the karaoke vocal line for each song while also playing the guitar part. Afterward, he told me about his experience playing acoustic guitar:

DAN: The things I do are, I don't play lead guitar, so I can only sort of do rhythms with strumming, mostly acoustic and occasionally little incredibly simple lead bits.

KM: So when you're singing and playing simultaneously in Rock Band, it's harder material than you could do on a real instrument?

DAN: Yes, although it's not necessarily—it's harder to keep track of the different places I have to put my fingers, but there's less—all the things like, you know, when you play an actual acoustic guitar, every time you're strumming, you have to make sure you're hitting the right strings, and you have to hold on to your pick and everything. So the way the notes are changing is more complicated [in *Rock Band*], but everything else is simplified, which actually makes it less stressful I think.[18]

▶ 3.4

Dan is a singer-songwriter who performs at open-mic nights. Learning to play complicated lead guitar parts is not a high priority for him. Playing *Rock Band* offers Dan a way to gain a more intimate understanding of lead parts without remaking himself as a rock guitarist.

Mike, a Brown undergraduate, does play lead electric guitar parts. I asked him what aspect of the game was the most fun for him.

MIKE: Probably the solos that aren't as complicated as in real life. When you play the solos in Guitar Hero, it's up-and-down, with-the-buttons. But when you're playing the real solo on a real guitar, it's a different level of complexity. And when I play it on Guitar Hero, I feel like, "Oh, I can actually play this—the solo sounds great, and I'm playing it!" But it's actually just a few buttons that you see on screen as opposed to actually playing the solo. That's probably the most fun, for me, I get out of Guitar Hero is playing the solos.

KM: Are these solos that you don't think that you ever would be able to play on real guitar, they're just too hard?

MIKE: Some of them, especially like Dragonforce ["Through the Fire and Flames," a song often used to represent peak difficulty level in *Guitar Hero*]. I can play parts of the solo, but I can never play it in its entirety or the song in its entirety.[19]

▶ 3.5

Mike told me he had learned some riffs and "certain little techniques" through *Guitar Hero* solos and translated them back to electric guitar. He had integrated *Guitar Hero* into a well-established learning approach for "self-taught" instrumentalists. As Jones and Willis observe, "The sense of empowerment achieved by being able to play an instrument and reproduce the sounds of a favorite record is a common starting point for young musicians" (1990:79).

Lauren, a graduate student with experience playing guitar, bass, and drums, didn't enjoy playing the guitar or bass parts in *Rock Band* because they felt like "just pushing buttons, like any video game"—too far removed from playing a stringed instrument. However, she loved playing the *Rock Band* drum controller:

It's really similar [to a real drum kit]. I think the coordination, you know, just like the basics, is pretty much the same. I think it's actually good practice, like if you want to learn how to play drums. I mean, obviously, you know, the sound is very different. It's easier because there's cues. But it's also—that also makes it harder in a way, because you don't have the freedom to just do what you think sounds right. You have to do it in the way the game tells you to.[20]

▶ 3.6

There is a much more transparent relationship between the drum controller and an actual drum kit than is the case for guitar. Like Lauren, many drummers told me that having to follow notation was the biggest challenge presented by *Rock Band*, since they usually learned songs by ear. Over time, though, they became proficient *Rock Band* sight-readers and could transfer drum parts learned from game transcriptions back to their real drum kits.

Chris Sanders, a St. Louis player who is a proficient guitarist with additional choir and piano experience, broke down the technical challenges of playing the guitar controller at the "expert" level into three skill sets:

(1) Learning to quickly "scan" the rapidly descending note cascades on the screen and mentally process them into the appropriate notes. . . . This includes deciphering the "code" and recognizing the following: (a) what note (button) or chord (buttons) was being called for; (b) whether single notes required strumming or whether they were "hammer-on" or "pull-off" notes that did not require a simultaneous strum; (c) and using their spacing and relationship to the other notes to accurately interpret their timing/rhythm in relation to the music.

(2) The next step in the process was learning to manually carry out the now-decoded instructions. "Expert" songs tend to possess numerous rapid chord changes or "flurries" of notes and trills during solos. . . . This part is still a challenge as there are songs with "odd" (read: infrequently used) chord transitions that I cannot always perform quickly or accurately enough, or crazy solo parts that would require hours of practice in order to master.

(3) . . . When playing songs on Expert, you will often encounter challenging solos or tricky sections where you will not be able to succeed in hitting the right notes. Therefore, in order to "pass" the song and move on to the next challenge, you start learning when to use (and when to hoard) the game's Overdrive/Star Power feature that helps prevent [you] from failing out. Similarly, once you can play the songs without fear of failing, in order to increase your score you need to think strategically on exactly when to deploy the Overdrive/Star Power in order to maximize your score. For example, given a set number of measures in the song, you would earn more points for multiple individual notes than you would for a lesser number of sustained notes. Similarly, you get more points for chords than for individual notes. So you learn to analyze the structure and "charting" of the song and time your use of Overdrive/Star Power for those sections with more potential points. (survey follow-up email correspondence)

Regardless of prior musical experience, players learn to cultivate and rely on internalized technical skills as they tackle increasingly difficult material. Shaun Scovil, a 30-year-old who works full-time running *Rock Band* bar nights around Boston and described himself as having "no musical talent whatsoever," explained this process:

SHAUN: I've just recently stepped up to playing [guitar] on expert, and I can't keep up with it. I mean it just flies by so fast. So I miss notes, but I do well enough. And even if there's a barrage of notes coming at me that I know I can't possibly keep up with, my fingers just start moving to try and keep up with it because I want to hear the song. And it's amazing what it does for your muscle memory, like your hands just suddenly take over and say okay, I know where the buttons are and I'm just gonna do this. . . .

KM: Are you playing drums on expert?

SHAUN: No, I'm only up to hard on the drums.

KM: I'm just transitioning to playing on hard on drums. It's so much harder than guitar.

SHAUN: It is tough. The foot pedal is what kills me.

KM: Doing those double kicks.

SHAUN: Yeah. I've actually turned a corner though, very recently, within the past week, where my foot, I don't consciously think about the foot. The foot pedal comes, and my foot just bounces on it, and I just focus on the other-color drums.[21]

▶ 3.7

Josh, an undergraduate with many years of jazz saxophone experience, initially told me that playing *Guitar Hero* and *Rock Band* felt completely different from playing other video games: "It's a different kind of high, a different kind of fun. . . . The high with Guitar Hero is, it *feels like* you can improv on it. It feels like you're doing your own thing, and when you move up a level from medium to hard, you notice all the differences and how much harder the song is to play." But later in the interview, in a discussion of muscle memory, he identified experiential similarities that connected first-person shooter games, *Guitar Hero*, and playing sax. He began by talking about playing *Halo*, a popular shooter game:

JOSH: You've got to log in those memories to your fingers. Your fingers gotta be, they gotta be quick and they've gotta memorize and they've gotta understand what's going on. When you tell it what to do, it's gotta do it. So that's why you play it over and over because then your hands kind of get into the mode, and then they understand where they're supposed to go, and that gets you better than other people, even though [you're all doing] the same thing.

▶ 3.8

Then he shifted to talking about playing *Guitar Hero*:

JOSH: That's a huge thing [in *Guitar Hero*] because you've got to know when it's coming up and what's next because you can't really see—when you're replaying it, you can't really see it that quick, and it's got to be basically like you actually understand already what's going to be next, the next set of moves. You've got to, since you can't see that, you've just got to know it. So definitely, the muscle memory is there.

KM: And does that compare to your experience with sax?

JOSH: Yeah, yeah because sometimes when you have a quick trill or something that you have to do, you can't really think about it. When you're going to solo and you're going to trill up your scale and it's a jazz scale, you've got to know to miss some of the notes so it doesn't sound really bad. So you've just got to, it's got to be there. You've got to understand what you're going to press.[22]

▶ 3.9

These accounts from Chris, Shaun, and Josh highlight the similarities between playing *any* video game and playing notated instrumental music. In both cases players have to master a particular technical interface—a game controller or an instrument—in order to realize specific, time-sensitive visual instructions in an environment that provides multisensory feedback. Both practices rely on "[the] corporealization of perception, the translation of perception into bodily movement" (Lahti 2003:164). And in both gameplay situations and musical situations, in order to play well, one must practice certain activities until they are internalized and don't require conscious attention. These basic affinities are heightened in *Guitar Hero* and *Rock Band,* in which high-level players must seamlessly integrate notation-reading, analytical listening, and technical mastery of their instruments. Once players achieve a certain level of competence, they can relax into their parts and direct their attention to the rest of the band. As an interviewee told me,

> What I like to do is scan out from my one little line and look at the other ones. You can see how the instruments break down at certain parts of the song. There will be a guitar solo. There will be a drum fill. The bass might have a groove section. That aids the listening in a way. Once I see [a track] become more active visually, my ears will go to that part aurally and I'll start to key in on what's happening. . . . If I'm sight-reading and making mistakes and my part drops out of the sound, then I become really flustered. It all kind of falls off from there. If I'm playing well, especially if it's the second time a chorus has come around or a verse and I've played through the same part before, I'll relax. I'll start to rock back and forth. I might make comments to the other players and check out their parts. I can kind of expect what is to come and the patterns that are going to arrive. I can prepare them in my fingers. I zoom out and look at the whole picture a little more when I'm playing well.[23]

LISTENING LIKE A MUSICIAN

Like four-hands piano transcriptions, *Guitar Hero* and *Rock Band* bring a concert performance repertoire into the home. One crucial difference is that in the era before recording, four-hands transcriptions and other small-ensemble arrangements were the only channels available to make symphonic works "acoustically accessible" at home and repeatable on demand (Christensen 1999:259). Ideally the parlor musicians would also have heard a work performed live with the full orchestration at some point. From then on, they could "use the transcription as an *aide-memoire*" (264), filling in the missing sonorities and textures inside their own heads. *Guitar Hero* and *Rock Band* players can seemingly skip this step; when they play the games' transcriptions, the original recorded track comes out of the speakers. But the sound coming out of the speakers is not necessarily the sound in a player's head. The games foster a creative, imaginative listening orientation in which players feel responsible for producing the music through moment-to-moment embodied engagement with the "inner time" of the song (Schutz 1964–67 [1951]).

In an essay on listening and improvisation, Vijay Iyer suggests that "music can be viewed as a consequence of active listening; it is, at some level, *through* informed listening that music is constructed. Placing the skillful listener in such an active role explodes the category of experiences that we call listening to music, because it allows the listener the improvisatory freedom to frame any moment or any experience as a musical one" (2008:285). *Guitar Hero* and *Rock Band* teach players how to be active, informed, skillful listeners; in other words, players learn to listen like musicians.

> I've learned to listen to music differently. Whereas before . . . I would typically listen to a song as a whole, I now find myself picking out the various individual instruments and assessing their particular contributions to the music. I'm thoroughly convinced that the method these games use to "reward" good play (playing the particular instrument's soundtrack when you do well, and cutting it off when you don't) is directly responsible for this increased sensitivity to the individual instrumentation of songs. (survey follow-up email correspondence)

> When I'm listening to my iPod on the bus now, I can hear the guitar riffs more, and I'm playing the notes with my fingers, like I would on Rock Band.[24]

> Playing the songs (particularly in Rock Band) has completely changed the way I listen to music and my appreciation of it. Playing drums in Rock Band made me realize that I never really paid attention to drum and rhythm elements in music. Getting experience like this adds new layers to my experience of listening to music. (Mont Boge, male, 18–21)

In the past, I just hated rock band music. But after I played this, I found that the beat pattern can be so complicated, even though it's just bam, bam, bam, it's so noisy—but underlying the music, there's some complicated piece that I can admire. So you could say that by playing this game, I respect rock music more.[25]

[Rock drum lines] are more complicated than I pictured them being. It's not simply that you establish a beat for a measure and then play that beat over and over again. There are fills—maybe not as exotic fills as there would be in jazz, but the third and fourth time you play something, there may be slight variations that I would not have looked for before, because I just don't think like a drummer.[26]

I find myself understanding the anthemic quality of [pop rock] songs far better than when I was going through music theory and dissecting their simplistic chord patterns. (anonymous male, 22–26)

Since I started playing this I have gotten more into the heavier rock and metal. . . . Before, I thought it was just a lot of strumming on the guitar, like, recklessly. But when you listen to it after having heard the notes specifically being broken down into five colors, you appreciate, like, oh, he's doing this with a bass player, and the drummer in the background is definitely keeping it together.[27]

Along with changing players' experience of songs and genres they already know, *Guitar Hero* and *Rock Band* offer players a distinctive way to experience new music, both during gameplay and when listening in other contexts. In assessing a new song, players ask themselves whether it is satisfying to play, rather than whether it sounds good or is consistent with their ideas about their own tastes. This is in some ways a return to an earlier state of affairs. As Michael Chanan writes of musical reception in the nineteenth century,

As long as an audience is made up of listeners who themselves play and sing, their listening also becomes, in crucial respects, an active process. Such listeners are able to adopt an attitude which bases its musical judgment neither on subjective emotion, nor on detached intellect, but on a third matrix: the practical knowledge of the instrument, or the voice, which guides both intellect and emotion in the moment of performance. (1994:27)

Like nineteenth-century amateurs hungry for new sheet music, today's game players are eager to try out new repertoire, and they are willing to pay for it. In fall 2010, Viacom announced that players had downloaded over 75 million tracks from the *Rock Band* Music Store, from a catalog of about 2,000 songs by 800 bands (Viacom 2010). (Most individual songs are priced at $1.99.) These sales figures have attracted the attention of artists and labels; Harmonix's Greg LoPiccolo noted that there is "a long list of acts who want to get their music into [*Rock Band*]" (DeRogatis and Kot 2009).[28] The games contrast sharply with other "digital discovery" channels for music marketing and distribution, which tend to build on a consumer's existing tastes or those of his or her affinity networks (e.g., the predictive algorithms that generate music recommendations on Amazon.com, in the iTunes store, or through streaming music services like Pandora and Last.fm; see Jennings 2007). Two-thirds of my survey respondents agreed with the statement "There are [*Guitar Hero/Rock Band*] songs I enjoy playing that I don't enjoy as a listener"—as Roland Barthes notes in his essay on *musica practica*, "The same composer can be minor if you listen to him, tremendous if you play him (even badly)" (1977 [1970]:149). But often the pleasure of playing also crosses over to listening: Three-quarters of survey respondents said the games had increased their appreciation for certain songs or genres.

When asked *how* these games changed their listening experience, players explained that the combination of reading notation and the physical act of playing a particular part (guitar, bass, drums) made them hear songs differently, including songs they had never played in the games. Virtually all of my interviewees reported "playing along" mentally as they listened to music, sometimes also playing air guitar or drums (with game-controller performance mechanics in mind) and/or visualizing appropriate on-screen notation. Blake Perkins, a survey respondent who began playing real guitar after winning an instrument in a *Guitar Hero* tournament, told me, "I do still think about how a lot of songs would be tabbed out [i.e., notated] on Guitar Hero. The only difference is now I think a lot more about how it is played on real guitar. Both sides of it really interest me, and with the more knowledge I gain about each one, the better I understand how guitar and Guitar Hero work" (follow-up email correspondence). This is a textbook case of what Paul Willis terms "symbolic creativity." As Willis writes, "Consumption of cultural commodities involves its own processes of production (symbolic work and creativity, grounded aesthetics) in a way that is not true for other commodities. In short, a pop song is not a steel ingot" (1990:132).

The link between heightened listening and the movement practices associated with the games brings to mind Simon Frith's observations about dancing, which he characterizes as "a form of enhanced listening" that generates "a heightened, more intense, above all more concentrated sense of the music" (1996:223). Like dance, *Guitar Hero* and *Rock Band* gameplay could be described as "movement which *draws attention to itself,* in the very act of ceding control to the music. . . . The dancer's technique, that which allows her body to do whatever the dance requires, is precisely that which allows her to forget about her body altogether and just think the music" (221). This sense of immersive engagement, commonly referred to as being "in the zone," "in the groove," or in a flow state, is common among dancers, athletes, musicians, and participants in many forms of play and ritual (Csikszentmihalyi 1991). It is also an explicit goal of virtually all video game design.[29]

For many musicians, the affective experience generated by physical engagement and heightened listening is a defining quality of genuine, successful music-making. Harris Berger vividly describes a rock guitarist's "entering into the music" through intense attention to "flows of sound phenomena," a musical experience that Berger characterizes as a "creative act of perception" and a "culturally specific style in the organization of attention" (Berger 1999:153–54, 28, 69). Many *Guitar Hero* and *Rock Band* players—including myself—organize their attention and channel their affective experience in exactly this way as they play their plastic instruments. As one player put it, "It's like if you were walking over some terrain and each note were some kind of slope or bump, and instead of being aware of exactly where your feet are at the moment, you have a view of the terrain."[30] Sitting at my *Rock Band* drum kit, sight-reading my way through a song I've never heard before, I suddenly "feel the form" of the piece (Berger 1999:163), get into a groove, and the drums seem to play themselves. Some readers may protest that in *Rock Band* the drums *are* playing themselves. Yet I'm always aware that the music would stop if my hands and kick-drum foot stopped moving. This knowledge makes my efforts feel authentic, and that feeling generates a sense of affinity with the drummer who recorded this track.

CREATIVITY AND CONFORMITY

Players can readily identify and discuss numerous musical aspects of their gameplay experience, from their physical engagement with the instrument controllers to their sight-reading skills and altered listening experiences. However, some also point to a major obstacle to feeling as though they are really making music: the apparent absence of musical creativity, originality, and authorship. As one survey respondent wrote, "I can appreciate the music on a far deeper level while I'm playing it in Rock Band, but I am absolutely shutting down the creative centers of my brain to do it. It's kind of like after you've composed a song for a group and now you're just rehearsing it to death until it becomes automatic" (anonymous male, 22–26).

There is a fundamental mismatch between these games' musical-production mechanics and their guiding musical aesthetic. Rock musicians don't perform from notation, and the guitar hero in particular is "a representative icon of individual creative expression" whose musical-genius status relies on apparent originality and spontaneity (Waksman 2001:124). Kevin, the undergraduate player who brought up the "Cheeto stains" stigma of *Guitar Hero* virtuosity, summed it up this way: "When you go see an actual musician perform . . . if you get the perfect musician, they are going to perform something differently every time, and it's going to be amazing every time. But if you get the perfect Rock Band player, it's going to be the same."[31] That kind of conformist perfectionism is the antithesis of authentic rock, and it drives the "just pushing buttons in time" assessment of these games.

However, not all players would agree that the games' ultimate objective is the production of identical performances. Such a claim privileges audio output, excluding players' physical performance techniques and affective experiences. Moreover, no two audio outputs are likely to be identical. Even if one only considers guitar or bass parts—disregarding *Rock Band*'s free drum fills and the karaoke option—the game-generated changes in crowd noise, shifts in the volume of different tracks, and players' use of the whammy bar and effects switch all have a significant impact on the audible performance.

Players often have complicated views about their own creativity and musicality, derived from their approaches to gameplay and their preexisting assumptions about the nature of authentic musical performance. In my online survey, I asked, "Do you feel creative when you play Guitar Hero/Rock Band?" The form required a yes-or-no answer and requested an explanation of that choice. Among my 547 survey respondents, answers to the creativity question were two-thirds negative. However, the responses entered in the "Please explain" box suggest that players employ many different parameters for assessing creativity. Consider this sample of responses:

Players who answered "no"

Nothing really creative about working to master playing a game someone else made (anonymous male, 18–21)

im not being creative i am just re-enacting someone else's work (Jake, 18–21)

Guitar Hero is more a matter of pressing the right sequence of buttons at the right time. There isn't the freedom to actually improvise anything. The only way to be creative is in how best to incorporate the furniture into my stage act. (Boxy, female, Ottawa, Canada, 27–30)

there is no creation involved in playing Guitar Hero (yet). it is essentially the same exercise as playing Simon Says, which, to be successful, requires players to reproduce without personal expression. that said, in social situations, the crowd will reward players for being entertaining above and beyond what the game environment strictly allows. (davidicus, male, Los Angeles, 36–40)

Players who answered "yes"

Yes I do, because I feel like I am actually the guitarist playing the songs; the guitarist who spent years of dedication to create such works of music. (anonymous male, 18–21)

I like to feel as if I am actually playing an instrument and making music. It's a good feeling when I play well. (Saundra Doyle, Mt. Pleasant, Michigan, 18–21)

I feel like I'm jumping into the artist in their time and playing along and maybe even feeling what it was to be that creative individual in their time. Plus it feels good to attempt to thrash along with their solos. I get too wrapped up in playing correctly to enhance the experience (formative musical experience was in the orchestra . . . used to sitting and playing), but I try a little. (iRone, male, 31–35)

I am a creative person . . . so I think this is a substitute for some of the craft type stuff that I would otherwise be doing. So, I guess I have to say yes, it does in a way make me feel like I am creating music, even if I'm only pressing the buttons when I'm supposed to. (Donna, over 50)

Yes and no. No because I don't attempt to learn showboating techniques (behind the head, eyes closed, etc.). Yes because my mind analyzes how to play certain segments the more I play them. Although I can learn new techniques here [on score-hero.com], ultimately it is up to the creative side of mind that will help me coordinate the finger patterns necessary to perform a difficult solo. I cannot comprehend how a number of people (maybe 50) are able to "tap" Jordan. (mrkuo, male, 27–30)

As one might expect, many of the "no" answers to the creativity question point to the fact that *Guitar Hero* players are playing someone else's compositions, reenacting someone else's specific performance, and have almost no control over the resulting sound, apart from playing around with tremolo by using the whammy bar on long notes or simply dropping notes by making mistakes. Like many orchestral musicians, they readily acknowledge that "certain aspects of what they do are not, in fact, very creative, and simply require them to reproduce musical information in as straightforward a manner as possible, unfiltered . . . by their own interpretive ideas" (Cottrell 2004:120). However, a lot of the players who answered "no" go on to use the "Please explain" box to discuss the creative aspects of physical performance techniques—not only "showboating" moves like playing behind the head or with one's eyes closed, but also the specific fingering patterns or alternative playing techniques required for the mastery of challenging passages. A subgenre of *Guitar Hero* YouTube videos is devoted to teaching fingering patterns, "tapping," and hammer-on techniques for certain songs.

Many players also point out the creativity of those who have made custom song charts and software hacks for inserting them into the games. Gabe, a survey respondent from Denver, answered "yes" to the creativity question and explained, "In Guitar Hero, I'm just hitting pre-placed notes. But when I make a custom song chart, I can inject my own style into it." A thriving area of ScoreHero.com (a player-produced *Guitar Hero* website) is devoted to these custom charts, and the chart makers also advertise their work on YouTube (e.g., backseatstuff 2007). Over 6,000 custom charts are available on Score-Hero alone. ScoreHero and other forums also track numerous hardware hacks, ranging from custom paint jobs, to adding more realistic weight to the controller by gluing stacks of metal washers inside its body, to rewiring the *Rock Band* guitar for use in the *Guitar Hero* games (FallenSeraph 2007, wumpus 2007, P4721cX 2008).

The discussions on these web forums demonstrate that guitar controllers can acquire the same collectible fetish-object status as real guitars. A *Guitar Player* writer made this connection explicit: "For a while, it even appeared that PlayStation, XBox, and Wii were supplanting Fender, Gibson, and PRS as objects of desire for young and old alike. But, ironically, it may turn out to be a video game that helps shift the balance back" (Ross 2008:59). Some players invest considerable creative energy in both individual instrument customization and generalizable modifications designed to bring the controller closer to a real guitar's size, look, and feel. Brian Shandra, a volunteer moderator for the ScoreHero.com player forums, has customized numerous controllers, including one that will shoot real flames (Shandra 2010; see Figure 3.6). He described his initial creative impulse this way:

> I said, you know, I have this black and white guitar like everybody else. It is an integral part of my playing experience: I pick it up, I touch it, it's my interface with the game. . . . What are the things to make this guitar special and unique, to make that game more interesting for me to play? . . . Having something so special in my hands does make me feel more connected to the music that I'm

Figure 3.6
Brian Shandra with one of his modified *Guitar Hero* controllers. Shandra posted this image on the ScoreHero.com forums; it circulated widely online and inspired numerous parodies. See Shandra (2010). (*Photograph courtesy of Brian Shandra*)

creating, so it does help complete the picture for me—to help create that experience and make it whole.[32]

Many of the survey respondents who answered "yes" to the creativity question gave explanations that had nothing to do with developing special techniques, showboating, making custom charts, or modifying controllers. Instead, they pointed to the creative aspects of role-playing—"jumping into the artist in their time," in iRone's words, or imagining another form of rock-star identity. Still others reported that *Guitar Hero* made them more creative listeners or that it amplified the creative inspiration that they usually got from listening to music. One player referred to a sense of "channeling" a creative force that already existed, as though he were stepping into the flow of music-always-already-in-progress. Steffen, the rock drummer who wrestled with the eerie "accomplished feeling" he derived from playing guitar riffs in *Rock Band*, told me, "I don't know necessarily if you are *being* creative, but you certainly *feel* creative."[33] When I asked Sean, an interviewee who plays classical piano, whether he thought these games included any elements of creativity, he told me that he had been thinking about this issue for a while:

SEAN: Well, first I thought, oh, that's dumb. Obviously it's not creative because you didn't think of it and you have to just do what it tells you. But it's—I guess I thought of "the dictionary definition of creation" and, you know, it's making something. I guess you are making something.

KM: In what sense are you making something?

SEAN: Well, you're causing a song that—I guess you're allowing an audio track to play by replicating it well enough. But—I guess the next step from that: it's not that or be Stravinsky, you can just play it like a Mozart sonata. And is that not creative because you didn't write it? That's the kind of grey area that I was thinking about.[34]

▶ 3.10

When players grapple with the creativity question, they are testing their gameplay experience against various received ideas. Is creativity an individual trait, as in "I'm a creative person"? In that case, if you say playing *Guitar Hero* makes you feel creative, you might mean that it allows you to pretend to be someone else: the original artist, already culturally endorsed as "creative." Or you might deny feeling creative when you play, because you identify yourself as a creative person but you don't think the games leave any room for expressing that creativity. As one player told me, "When these games are played perfectly, it will sound just like the original recording. So you don't have that creative input that any sort of live performance will give you."[35]

Both of these responses assume that creativity is verifiable through its tangible products. Thus, as Sean explained, *Guitar Hero* might count as creative if everyone could agree that players are "making something . . . causing a song." Sean's more specific model of "just playing it like a Mozart sonata" suggests a further criterion: When classical musicians are credited with creativity, it's because of their individual interpretation of existing works. As Stephen Cottrell puts it, "For most [classical] musicians . . . being creative is essentially synonymous with having the opportunity to express musical individuality, with feeling that you are putting something of yourself into the performance" (2004:113). Many players say that *Guitar Hero* and *Rock Band* simply don't allow for this kind of creativity, apart from drum fills and the guitar's whammy bar.

However, this assessment assumes that a player's interpretation has to be sonically realized and verified as "creative" by a listener. Consider the two dimensions of Cottrell's definition: "expressing" musical individuality and "feeling" that you are putting yourself into the performance. Suppose that you recognize the feeling even when the actual expression is absent? What if you conceive of creativity as an internal orientation or "stance," a particular quality of engagement with what you're doing (Berger 2009)? Then you might approach the creativity question by considering whether playing these games tends to move you into that creative orientation, regardless of whether you've invented any new riffs or generated any sounds. Josh, the jazz saxophonist, seemed to be invoking this concept of creativity when he said, "The high with Guitar Hero is, it *feels like* you can improv on it. It feels like you're doing your own thing."

If players adopt a creative stance as they play the games' repertoire, if they can imagine that they are having a real-time interpretive impact on the songs coming out of the speakers, does it matter that they aren't really generating sounds? After all, in the course of performance, no musician ever experiences his or her own sonic output in the same way as a listening audience member.

CONCLUSION: IN THE SHOES OF A ROCK STAR

The unconscious, fluid, internalized skills cultivated by *Guitar Hero* and *Rock Band* players help them to create an ever-closer correspondence between their own movements, the on-screen notation, and the music coming from the speakers. The more they perfect this correspondence, the more satisfying the gameplay experience—and the more troubled some critics become. Brian Shandra, the ScoreHero.com moderator who can shoot flames out of his customized guitar controller, explained:

> I think the fact that it does strike a chord with so many people in that way, both in a positive and negative way, really speaks to how visceral the experience is and

how much more—it's not really just a game anymore. Since you're holding a plastic guitar, you're moving your body, you're getting into it, it almost puts it into a very grey area between, is it a game, or are you actually performing? That grey area, people are sometimes uncomfortable with.[36]

▶ 3.11

My experience with players and with media discourse suggests that Brian is right to associate that uncomfortable "grey area" with the physicality of *Guitar Hero* and *Rock Band* performance. When players feel like they're "making something" or "causing a song," that feeling is intimately linked to their embodied experience of gameplay (cf. Blaine 2005:32). As I discussed in the *Grand Theft Auto* chapters, there is typically a huge gap between the embodied experience of a video game player and the actions of his or her on-screen character: In the GTA games you can press one button on the controller, and your avatar might do something as complex as running up to a moving car, opening the door, throwing the driver out onto the pavement, and hopping into the driver's seat. But *Guitar Hero* and *Rock Band* are designed to encourage players to work toward narrowing that gap. It's as though GTA gameplay revolved around gradually training players to move from pushing one button to acting out the challenging physical maneuvers of a carjacking.

By giving players an immersive gaming experience filled with rock-oriented cues—including the musical repertoire, archetypal rock-star avatars, a responsive crowd, a guitar-shaped controller, and physical performance hints—*Guitar Hero* and *Rock Band* encourage players to adopt a rock performer's identity. As James Gee has noted in his work on gaming and learning, "When learners adopt and practice such an identity and engage in the forms of talk and action connected to it, facts come free—they are learned as part and parcel of being a certain sort of person needing to do certain sorts of things" (2006:176). In the case of *Guitar Hero* and *Rock Band,* the facts that come free include new modes of musical listening, a sense of the physical relationship musicians develop with their instruments, an intimate knowledge of a particular selection of songs, and assorted elements of rock history and ideology (some of which are presented in explicitly didactic form, written on a chalkboard that appears between songs). In response, many players are closing the gap between virtual and visceral performance not only by achieving a note-by-note correspondence between physical input and audible output, but by performing like rock stars: They engage physically with their musical performances in ways that go far beyond what is required to provide accurate inputs to the game software.

In the fall of 2007 I interviewed Rob Kay, the lead designer for *Rock Band* and the original *Guitar Hero* game, in his office at Harmonix Music Systems. At the time of my visit the company was on the verge of releasing *Rock Band.* Kay and I discussed the original design intentions for *Guitar Hero,* including how designers worked toward "bringing people the feeling of playing music."[37] He told me that *Guitar Hero* was never meant to be a guitar simulator. Instead, it was intended to be a *musical-performance* simulator and, more specifically, a rock-performance simulator:

> We definitely wanted, in the design stage, to find a way to make the simulation more than just a clinical recreation of music, and we had this whole kind of ambition that started out pretty loosely articulated to bring some showmanship, as well as musicianship, to the experience. So we decided early on that we were going to have this metric of "star power," and we didn't know what it was. . . . Somehow, we were going to try to find a way of raising your rock-star persona and your showmanship.

The solution devised by the design team was an on-screen energy meter that fills up with light when a player performs well—which means not only hitting a lot of the notes, but also using the guitar controller's whammy bar on sustained notes and accurately playing complete musical phrases. The game presents this meter as a representation of crowd approval. Anytime after the meter has passed the halfway point, the player can deploy "star power" by raising the neck of the guitar controller, which has an integrated tilt sensor. The result is a burst of multisensory feedback: The screen explodes with light and sparks, the notation track shakes, the crowd roars, and all the notes turn electric blue. (The color change provides an additional spur to performance adrenaline, because it requires the player to read notes by their relative horizontal position without the aid of color-coding.)

In an article about designing new interfaces for musical experience, Blaine and Fels note that "While the affordances of the sensors and interface should be transparent to the players, understanding their individual impact on the system is critical. This can be achieved through the use of music, lights, images, sound effects, or a broad range of other possibilities; anything that supports the intentions of the players will serve to reinforce the perception of a highly responsive system" (2003:133). Brian Shandra's account of his first encounter with the "star power" tilt sensor is a testament to Harmonix's success at integrating this feature into the overriding design of the game:

> Funny thing, when I first picked up the game, I did not understand that part of the game mechanics. I found it odd that when I was really into playing the song, I naturally did that. . . . I was like, why is the screen going all blue when I'm really feeling it? I didn't understand—I wasn't purposefully tilting it to get that reaction; that was my natural thing to do. I was really rocking, I tilted that guitar up, and the crowd and the screen reacted. I know it's digital, it's all

computer-generated response, but in a way, I felt that my own emotional response to the song was being mirrored on the screen.[38]

▶ 3.12

The game designers at Harmonix were very much attuned to the emotional impact of this programmed feedback, and to the significance of requiring an iconic rock-performance gesture—raising the guitar neck—as part of successful gameplay. Rob Kay told me,

> I think we were always kind of keen to get people doing that move, because it's really obvious that once you give people this cue that their physical performance has got something to do with the game, even though that's the *only* physical performance that has something to do with the game—you're suddenly in the mindset and thinking about all of those rock-star moves that you see people do, and then they will just jump around and do the rest. And it seems that that's often the case with these kinds of things. You don't have to go the whole way. You just need to give people a beginning—a shove in the right direction, and they'll do everything else.

Brian Shandra and Rob Kay both point out that rock-performance moves seem to come naturally to players; all it takes is a "shove in the right direction" provided by the games (cf. Isbister and Höök 2009:4). But of course we know there's nothing really "natural" about this phenomenon. Instead, players' performances recall Blacking's dictum that "music has no effect on the body or consequences for social action, unless its sounds and circumstances can be related to a coherent set of ideas about self and other and bodily feelings" (1995:176). *Guitar Hero* and *Rock Band* are so successful because they rely on players' preexisting knowledge and embodied understanding of rock performance. As Rob Kay told me, "Even just the process of playing, playing a guitar and holding something, and kind of doing this with your right hand, and doing this with your left hand. People recognize that motion from guitarists, and suddenly they're less playing a game and pretending to be a rock star, and more putting themselves in the shoes of a rock star." The designers also chose to enhance *Guitar Hero*'s symbolic cohesiveness by sticking with a rock-oriented aesthetic and repertoire. Kay explained why they didn't include other guitar music, such as blues or country, in the first version of the game:

> It just seemed right to us to make it about, you know, balls-out rock guitar. . . . It's the iconic thing. . . . I was always joking [with other designers], "It's got to be like a balls-out American thing." People identify with that I think, and it's just the clichéd thing, and often in video games it's those clichés that are easy to hold onto and get into. You just—you step out of yourself for a little while. And I

enjoy it even though I wouldn't necessarily identify, for me, that American guitar rock as being my favorite thing. I know what it is and I know how to get into it.

It's important to bear in mind that these games are not only drawing on rock's gestural repertoire, but are now playing a role in transmitting that repertoire to new practitioners—like the children who are having their first encounters with 1970s and '80s rock and metal through the games. As Tomie Hahn notes in her work on traditional Japanese dance, "Transmission systems are valuable to observe as processes of embodiment, effectuated via the senses, that encode and convey cultural meaning to reveal a particular (sub)culture's sensual orientation in the world" (2007:3). Playing these games is a form of "learning through the body," and that learning inculcates what Hahn calls "sensational knowledge" of particular aesthetic ideals (171; cf. Willis 2000). Here, too, we might come back to Blacking, who suggested that to understand Debussy's music, you ought to consider how he held his hands and body at the piano—and that you should "find out by feeling for it" as you sit at the keyboard yourself (1973:111). Blacking argued more generally that cross-cultural or transhistorical musical understanding always relies on embodied experience; "to feel with the body is probably as close as anyone can ever get to resonating with another person" (ibid.). Perhaps this is what Rob Kay was getting at when he made the distinction between pretending to be a rock star and putting yourself in the shoes of a rock star.

This distinction is important, since contrary to popular belief—and even contrary to some specific design goals—the satisfaction players derive from these games is not always related to fantasies of stardom. Among most players I encountered, the games' capacity to inspire the visceral engagement of playing music seemed far more compelling than the opportunity to enjoy adulation from a star-struck virtual audience. Consider this exchange with an interviewee:

KM: I interviewed Rob Kay, who was the lead designer for the first Guitar Hero and for Rock Band, and he said that they weren't trying to make a music simulator, but that they were trying to imitate the feeling of playing music. Does that distinction make sense to you?

KEVIN: Yeah, I think so. . . . You're holding a guitar like a guitar; you're strumming like you would a guitar, kind of; and you're hitting the buttons like they were frets or something. And, yeah, it *feels* like you are doing something as far as making music.

KM: It's interesting that you would go in that physical direction because I thought Rob was talking about the rush of performing, the emotional feeling of playing music.

KEVIN: Well, I don't know. I have never felt—I think when I play, it's kind of a personal feeling for me. I feel like I am making this for myself. I want to do well because I want it to sound good for me. And if there are other people around, that's nice, too. But it's always been personal for me, a way of relaxing. It's never been about performing for me.[39]

▶ 3.13

Of course, some players do value public performance, whether it be for actual, virtual, or imagined audiences (see Chapter Four). But people like Kevin offer a reminder that before we make assumptions about players' fantasy lives or performance aspirations, we ought to consider how these games also foster the pleasures of musica practica: everyday amateur music-making at home. Players who adopt the physical gestures of rock performance may do so simply because they know that's how rock music is supposed to be played, not because they are imagining a screaming crowd.

When people wonder why *Guitar Hero* players don't play real guitars, they neglect to acknowledge the possibility that these games might be compelling and valuable not just as simulations, fantasy enablers, or stepping-stones to real instruments, but because they offer people a new musical experience that complements and informs their other musical practices. *Guitar Hero* and *Rock Band* require players to read notation and translate it into instrumental technique. They generate the thrill of sight-reading through an unfamiliar piece of music, feeling it unfold under your fingers, and hearing it move in surprising directions, all the while requiring you to stay absolutely focused and locked into the tempo of the whole band—"a cross between riding a roller coaster and walking a tightrope," to borrow Christensen's description of playing a four-hands piano arrangement (1999:294). The games encourage repeat performances of the same pieces, which afford analytical insights into song structure and instrumental technique. They teach players to listen like musicians, whether they're playing through a new song or listening to the radio. In short, playing these games "feels like" making music to so many players not because of some sort of false consciousness or cult of repetition, but because both the games and the players are very musical—whether or not anyone ever picks up a real guitar.

CHAPTER 4

Just Add Performance

Staging Schizophonia

A video camera pans over the wheels and body of a red motorcycle and comes to rest on its rider, who introduces himself as he dismounts: "What's up, Internet? My name's Freddie." As a deferential roadie helps remove his black leather jacket, Freddie Wong proclaims his intention "to come and rock *you* with 'YYZ.'" He explains the heavy chains he wears around his neck: "The reason I have them on is that my solos are so blisteringly fast that if I didn't keep them tied down somehow, I might impregnate women." Another assistant hangs a *Guitar Hero* guitar controller around Freddie's neck, and he turns to a TV topped with liquor bottles to begin a virtuosic rendition of Rush's "YYZ." *Guitar Hero II*'s on-screen streaming notation is superimposed over close-up views of his performance. Every hammer-on and pull-off has its exaggerated flourish; Freddie plays some passages with the guitar held behind his head, turns away from the screen to demonstrate his mastery of the material, and often lifts the guitar neck into rock-god phallic position. At the end of the song, he smashes his instrument into pieces. This YouTube video was viewed over 8 million times between October 2006 and April 2011 (freddiew 2006; see Figure 4.1, ⏵ 4.1). When he made the video, Freddie Wong was a film student at the University of Southern California. He took first place in the 2007 World Series of Video Games *Guitar Hero II* competition.[1]

In this chapter I address public *Guitar Hero* and *Rock Band* performance, carried out at tournaments, at bar nights, and on YouTube. These games are deeply theatrical, by design. As Samuel Weber notes, "placement" is a defining trait of all theater: "the arrangement of the place, the positioning of the people and things in it, is constitutive of what is taking place there" (Weber 2004:4). Like all

Figure 4.1
Still from Freddie Wong's "YYZ" video (freddiew 2006). (*Screenshot by the author*)

video games, *Guitar Hero* gathers players (and appreciative audience members) around a screen; their gameplay is always configured in a particular space. Players rely on special props and stage machinery—the game-controller instruments and game console—and often employ a recognizable repertoire of rock-star moves in the course of their performances. As they play, their on-screen avatars perform this same gestural repertoire in a second theater, a virtual space that has been meticulously designed to mimic a live rock performance venue: A crowd screams or boos, feedback crackles through the speakers, elaborate lighting flashes across the stage. (The games feature a wide range of performance venues, most modeled on identifiable real-life clubs or arenas.) When people play these games in bars or at public competitions, the avatar's stage and the player's stage sometimes mirror one another; at other times, the juxtaposition highlights the difference between the two performance venues and their respective performers and audiences.

This basic distinction between virtual and actual staging grounds is crucial in the analysis of any digital gameplay experience. However, the performance frames for *Guitar Hero* and *Rock Band* can be far more complex than this binary opposition might suggest. In the Freddie Wong video, for instance, at least six performances are layered together: the edited video performance designed for YouTube, the living room performance, the avatar band's on-screen performance, the human performances in a motion-capture studio that provided the physical model for the avatar band, a studio band's cover performance of the song "YYZ" (the game-development company wasn't able to license the Rush recording), and Rush's studio recording of the song. Of these six performances, I would argue that it is Freddie's energetic physical performance for an audience of friends in a living room that comes closest to the rock ideal of an authentic live show—yet it's his performance that is also the most explicitly theatrical, a parody of rock authenticity.

Spectacular physical performance is the most obvious, recognizable creative contribution that players can make to their gameplay. Physical creativity can be verified and acknowledged by an audience, unlike the internal orientation of "feeling creative" that I discussed in the preceding chapter. It is also something that goes mostly unrewarded in terms of earning game points. This fact has contributed to the formation of two broad camps of players, which I will call the *score-oriented* and the *rock-oriented*. Members of both groups are generally *performance*-oriented, but they employ different performance-evaluation criteria. As *Guitar Hero/Rock Band* lead designer Rob Kay put it,

> I'm all about the showmanship, when I play, and I think that that's what it should be really about if you're playing in front of other people. . . . They're both valid but in totally different ways. . . . The Score Hero guys [i.e., the gamer community at ScoreHero.com] are amazing. . . . They figured out that when you deploy [the high-scoring "star power" mode] at the exact right time you can squeeze an extra gem out of the deployment, and they know how many of those squeezable gems there are in every song.[2] When somebody's got an incredibly high score with 100 percent notes hit and they seem to have deployed star power in all the right places, they still know how to get the extra couple of points by deploying a fraction of a second later. It's kind of insane I think. I'm really impressed by that. The nerd in me is really impressed by that. And then, I guess the human in me is more impressed by people who [play the games] for showmanship, for complete performance.[3]

Some players use blended criteria to assess what constitutes impressive gameplay, as implied in Rob Kay's description of his own "nerd" and "human" aspects. However, the broad distinction between score-oriented and rock-oriented players offers a useful framework for considering players' motivations and performance strategies—and one that contrasts productively with the game industry's common distinction between "casual" and "serious" players. Consider how these terms are used in online promotional copy for *Rock Band 3*: "For the casual player, *Rock Band 3* has great new party modes that allow fans to get rocking with friends quicker than ever, including party shuffle and persistent drop-in/drop-out and difficulty selection from any gameplay screen. For the serious player, the revamped Career Mode features 700+ goals and rewards and seamless leaderboard integration for an endlessly deep campaign experience" (Harmonix Music Systems 2010b). This description encourages all potential consumers to sort themselves into one of two groups—either stereotypical "gamers" or people who like having fun with their friends—while also asserting that *Rock Band 3* is versatile enough to meet the distinct needs of each. Note how "casual" indexes social gameplay, music fandom, parties (i.e., face-to-face interaction), friendly/collaborative "rocking," limited technical competence, and short play episodes ("drop-in/drop-out" allows a player to join or leave a band in the middle of a song). "Serious" indexes in-game goals/rewards, a completist mindset (can you achieve all 700 goals?), technical expertise, online competition with public documentation of player rank ("leaderboard integration"), immersion ("endlessly deep"), and a war/adventure game paradigm, in which one embarks on a heroic "campaign" to defeat a series of increasingly difficult challenges. But as the examples in this chapter will demonstrate, both score-oriented and rock-oriented players identify as music fans, build collaborative *and* competitive social relationships through gaming, and often demonstrate a sustained investment in their gameplay that is far from "casual."

The score-oriented treat these games as well-defined rule-bound systems, in which the main challenge and satisfaction lie in determining how to exploit the scoring mechanism to best advantage. They strive to be "score heroes," and they share tips, techniques, and evidence of their triumphs on websites like ScoreHero.com. Their form of schizophonic virtuosity foregrounds speed, dexterity, efficiency (no extraneous body movement), total mental focus, and strategic innovation. Score-oriented players point out that "star power" can be activated using a button on the guitar controller rather than the tilt sensor; they see no reason to risk accuracy by changing their physical position. They generally don't feel creative when they play, and they often indicate that the question irritates them: The games aren't designed to reward creativity, so why make an issue of it?

Rock-oriented players have quite different ideas about the purpose and potential of these games. Their theatrical performances represent glam rock's legacy. As Auslander notes, in the 1960s "the counterculture's deep investment in the idea of authenticity entailed a necessary antipathy to theatricality . . . derived from . . . the emphasis on spontaneity and living in the present moment, the desire for community, and the suspicion that spectacle served the interests of the social and political status quo" (2006:10). In the 1970s, glam performers began to undermine those ideals by "present[ing] clearly staged spectacles [that] opened a gap between the figures on the stage and the 'real' people performing them" (37–38). *Guitar Hero* and *Rock Band* dramatize and restage this "opening of a gap" by reframing the virtuosic rock repertoire in a glam performance mode. Auslander's account of a televised, lip-synced 1972 Marc Bolan performance shows how closely rock-oriented players follow in the footsteps of their glam forebears: "Bolan took a vocabulary of movement and gesture that is associated in musical performance with the physical expression of musical sound structure, and the emotions attendant on playing, and used it essentially as an autonomous choreographic language from which to build physical routines" (96–97).

The spectacular schizophonic performances of rock-oriented players share this choreographed yet affect-driven quality. Rock-oriented players recognize that rock authenticity is performative. They generally do value their video game high scores, but they also believe creative performance is its own reward. Rock-oriented players aren't all extreme showboaters, but they play guitar standing up, they love Freddie Wong's videos, and they would never dream of activating "star power" by discreetly pressing a button. As they play these games, they explore the implications of their role as live performers of prerecorded songs.

A NIGHT WITH THE ROCK-ORIENTED: FOUR-PLAYER KARAOKE?

When I arrived at Our House East, a college-crowd Boston bar near Northeastern University, I found Mike Dadmun, the guy who ran the weekly *Rock Band* game night, standing out front tapping away at his cell phone—texting me, as it turned out. I recognized him from his Facebook profile photo and introduced myself. (I'd found him through a *Rock Band* group on Facebook and had messaged him to arrange an interview.) We went inside, where he had set up the *Rock Band* instrument controllers facing a large wall-mounted TV. It was only 5:30, and the college students were gone for the summer; the small bar was almost empty. The setup in the bar wasn't quite what I had expected. I had assumed the players would face an audience, as at a karaoke night—perhaps even standing on some kind of stage, with a projector screen across the room so that they could read the game's musical notation. Instead, the instruments were only a few feet from the TV; the players would have to look in that direction, with their backs to the rest of the bar.

Mike offered me a homemade brownie from a Tupperware container and insisted on buying me a beer. He was a super-friendly guy, and I could immediately see why he'd been hired to run an event that required people to volunteer to perform in front of strangers (see Figure 4.2). As people began to trickle into the bar, Mike seemed to know almost everyone. He told me that a core group of regulars had been coming to various *Rock Band* bar nights around town several times a week for months on end. The bars paid Mike and his employer, Shaun Scovil, to bring in the game equipment and host these events, which attracted customers on slow nights of the week. In the summer of 2009 their two-person company was running eight weekly *Rock Band* events around Boston. Similar bar events had become popular around the country in the wake of *Guitar Hero II*'s commercial success; a 2007 *New York Times* feature dubbed *Guitar Hero* night "the new karaoke night—without the embarrassment of atrocious vocals" (Zezima 2007). When *Rock Band* was released later

Figure 4.2
Mike Dadmun at a *Rock Band* bar night. (*Photograph courtesy of Shaun Scovil*)

that year, karaoke vocals became an option at game nights, though most participants still preferred to play instruments.

Mike described the crowds at the Boston bar nights as more diverse than a typical "karaoke demographic," in terms of age, gender, and musical tastes. (In Mike's experience, the local "karaoke demographic" comprised mostly women in their 20s and 30s who wanted to sing pop songs.)

MIKE: I think with Rock Band, it relates to a totally different demographic because it's more based on the rock music [included] when it first came out. As [the series] progressed, they introduced a lot more pop songs and whatnot. I think it gets more people involved. . . . There'll be age ranges from 21 to 60. And you get 60-year-olds up there playing "Paranoid" [by Black Sabbath] or something and wanting to sing it. And the way that we get them involved is to talk to them about it, make them feel comfortable about it, and show them what to do. . . .

KM: Do you see people being self-conscious about it?

MIKE: They are. A lot of people are self-conscious. I think everybody's self-conscious in a lot of different ways, I mean, doing something that you're not used to doing. But I think liquid courage really helps them. I encourage people, but I don't force it on them. I'll let them know how much fun it is, I'll want them to know it's in no-fail mode [a setting that lets the players finish a song no matter how badly they are doing in terms of hitting the right notes]. I'll give them a song list, and I go, "Check it out, there's 650 songs. Just watch people and what they're doing, and if you guys feel like you wanna get involved, go for it."[4]

▶ 4.2

Despite having significant musical performance experience, I had been nervous about the prospect of playing *Rock Band* in public. I normally played guitar on the game's "hard" setting and drums on "medium" or "hard," depending on the song; I was afraid that the bar-night regulars would all play on "expert" and might be scornful of those less accomplished. But after Mike's pep talk during the interview, I sat down at the drums right away; others picked up the bass, guitar, and microphone. We introduced ourselves to each other, settled on a song that the singer knew and none of the instrumentalists vetoed, and began to play. I was sight-reading the drum part; as the notation began to scroll down the screen and I saw the impending double-kick patterns, I experienced a heady adrenaline rush that made me play ahead of the beat for the first several measures. Then I got a sense of how those patterns would repeat and vary, and I relaxed into the tempo of the song. I could hear the sound of my sticks on the drum pads align with the steady click of the guitarist's fret buttons and strum bar. By the time we got to the second chorus I could look away from my notation track for about four bars at a time, which let me check out the other players' parts on the screen, watch the singer dance to the music, think about how to vary the basic beat in my next drum fill, and trade occasional words of encouragement or sympathy with bandmates: "That solo was brutal!" At the end of the song, we congratulated each other and passed our instruments to the next group. When our scores came up on screen, they showed that the band members had ranged from 63 percent to 99 percent accuracy in playing their parts, with each person playing at his or her own preferred difficulty level. No one paid any attention to the game points (calculated differently than the accuracy percentage). The singer congratulated the guitarist on his 99 percent score and said, "Hey, I bet you could play that on expert!"

As groups of four took turns, the rest of the crowd chatted and ordered drinks. Some bar patrons—especially those waiting to play—paid enough attention to shout praise, sing along or lip-sync to the occasional chorus, and clap at the end of each song. Others ignored the game entirely, treating it like a pool table or pinball machine—though the music was turned up loud enough to require near-shouted conversation. Most players seemed to keep their eyes focused on the screen, but they also got physically involved with their performances. Guitarists played standing up and moved to the music, and drummers often twirled their sticks or tapped out the tempo at the start of songs. Singers were the least dependent on the game notation (in part because they were often accorded the privilege of choosing the song), so they had the most freedom to turn away from the screen and interact with bandmates, friends, or audience members. For the most part, these weren't showboating, spectacular performances like the ones that circulate on YouTube, but they were physically and musically engaged. (See Figures 4.3 and 4.4.)

Figure 4.3
Two *Rock Band* bar-night participants. (*Photograph courtesy of Michael Dadmun*)

When Shaun and Mike present the *Rock Band* bar night concept to bar owners or coax bar patrons to play, karaoke is the obvious point of reference. Indeed, karaoke's particular form of "mediated-and-live" musicality (Keil 1984) informed the design of *Guitar Hero*. Harmonix developed a game called *Karaoke Revolution* (2003) before creating *Guitar Hero* and *Rock Band*. Harmonix designer Rob Kay told me this game "was very much designed with the explicit goal of going for your nontypical gamers. . . . Konami came to Harmonix and were, like, 'Oh, we want to do a karaoke game, and we specifically want to go out and get the people that don't normally play games. We want to get a lot of women to be playing this game.'"[5] The game did not sell very well. Subsequently, *Guitar Hero* was designed to appeal to the traditional young, male gamer demographic (in large part through its focus on the electric guitar and the guitar-driven rock repertoire) as well as to older/younger/female players, spanning what Cornelia Brunner calls the "butch–femme continuum" of game styles (2008). With *Rock Band*, Harmonix built on *Guitar Hero*'s success by expanding the range of available musical styles (including more pop songs and female vocalists), reintroducing the singing component developed for *Karaoke Revolution*, adding incentives for collaborative

Figure 4.4
Rock Band bar band. (*Photograph courtesy of Shaun Scovil*)

ensemble interaction, and retaining the technically challenging instrumental parts that made the games appealing to experienced gamers and/or people who "don't sing." In public gameplay contexts like Shaun and Mike's bar nights, *Rock Band* both invokes and recalibrates karaoke's "delicate balance between individuality and communality, and between imitation and invention" (Fornäs 1998:134).

Shaun told me that when he recruits players at bar nights, "I never refer to it as a video game. People who object to playing it, sometimes they'll say, 'No, I don't like video games.' I say, 'Well, I don't either. This is not a video game. This is more like four-player karaoke.'" However, he hastened to add, "I'm quick to differentiate it from karaoke in that if they're the one singing, they're not going to have to hear their own voice if they don't want to." Built-in volume controls let the player or the event organizer adjust the vocal mix coming out of the speakers. As Mike explained,

> I'll be there right next to them, increasing either the microphone volume or the vocal track, or lowering it. So I ask somebody, do you want to hear yourself? Do you want the lead singer of the vocal track to overpower your voice, or vice versa? . . . And I let them know, it's in no-fail mode. If you can really sing, who cares about the score. It doesn't matter. Everyone's gonna hear you, everyone's gonna know.[6]

Thus *Rock Band* can appeal to experienced karaoke singers—people who "want to have their own songs and like the attention factor of it," as Mike put it—while also being accessible to those who lack confidence in their vocal abilities. Meanwhile, potential new players who "don't like video games" can be lured to try the instrument controllers with the promise that it's just like karaoke, except they'll have the support of a group and don't have to sing—or even face the audience. Mike noted, "If you look at karaoke, people are on stage. They're right in the spotlight. With Rock Band, you're actually staring at a screen, so they don't feel as intimidated about being watched or on stage. And they're really not concerned with what people are thinking of them when they're playing. They're just more concerned about understanding what to do."

Shaun and Mike are experts at exploiting the inherent ambiguity of *Rock Band*—the question of "Is it a game, or are you actually performing?", as Brian Shandra put it (see Chapter Three). They present *Rock Band* as a performance opportunity for those who like to perform and as a low-pressure, social game for those who don't. Moreover, they point out that *Rock Band* can offer a *musical* performance opportunity for people who enjoy performing but believe that they "aren't musical" or "can't sing." Shaun put himself in this category:

> For somebody like me who has no musical talent whatsoever, it's as close as I'm ever going to come to playing a show at a bar, and I can just really get into it and let loose, and just have a good time. . . . Some of the bars that I've gone out to with karaoke, it's like people will only talk to you there if you're also going to sing. They appreciate your talent, and if you have something in common there, then they'll congratulate you and they might start talking to you. But with Rock Band, it doesn't matter. I've seen people just completely fail at the game, get zero percent of the notes, but they put on such a good performance—just acting silly and ripping away on the guitar—that people love them.

As event organizers, Shaun and Mike were invested in encouraging this kind of performance (see Figure 4.5). Shaun told me,

> I try to lead by example sometimes. It's a little bit outside of my comfort zone to do that, but I've seen the effects of it. . . . With the drums, some people, you can really see that they're getting into it, their bodies are moving, they're really feeling the rhythm. Even if they're missing notes, they have some kind of a rhythm going into it. The singer is usually the one who will have the most freedom to move around and really perform, get the crowd into it. Sometimes they'll get everybody clapping and singing along. Sometimes one of the guitar players will come over and sing along with the singer, and it just really depends on how comfortable everybody is with each other and how well they know the song, or how much they like the song. You can always tell when everybody in a room loves the song.

Like karaoke singers, *Rock Band* performers "can express themselves through song and announce their devotion to particular artists and genres, or they can momentarily transcend themselves by singing against type. There is the potential

Figure 4.5
Shaun Scovil plays *Rock Band* drums. (*Photograph courtesy of Michael Dadmun*)

not only for display, but for fantasy and epiphany" (Drew 2001:23). Mike told me, "It really opens your eyes to see things differently, when you see somebody of a different demographic playing a song. Like my friend Kellie: number one, a female, number two, an African American girl, singing 'Battery' by Metallica. I fell in love with Kellie the first night I saw her play that." But *Rock Band* players can also enact these kinds of performances solely through physical engagement, without subjecting their voices to listeners' judgments. Brian Shandra described it as a heightened form of choosing a song on a bar's jukebox:

> It bridges the gap between somebody picking out a song on a jukebox—"Oh, everybody's appreciating the music that I picked"—and actual performing of music. . . . When performing live, you don't necessarily have to stand there. I've seen a lot of people, and I've done it myself, dance around a little bit and get down on their knees and grind, play behind their back, all sorts of crazy playing styles that people will add to make it more of a performance.[7]

On my first night at Our House East I interviewed Mike's friend Kellie and her friend Heather, another bar-night regular. They were in their mid-20s and had gone to college together at Northeastern. Neither of them owned the game; they only played at the bar nights and generally stuck to the "easy" or "medium" difficulty level. Kellie had substantial karaoke experience; Heather had sung karaoke once or twice but didn't enjoy it. I asked them how karaoke nights compared to the *Rock Band* nights.

KELLIE: It's just a completely different experience. It's not better or worse than Rock Band, it's just different.

HEATHER: I think I like Rock Band better. I think karaoke's bigger and more out of control, and Rock Band is kinda more laid back.

KELLIE: Yeah, it's just relaxed. Maybe you come from work, and you just want to blow off some steam, you just maybe hop on the drums.

HEATHER: It's a good way to talk to different people, too. In karaoke, it's just you or you and somebody else on stage singing.

KELLIE: Getting drunk and singing.

HEATHER: It's not like you're communicating with the rest of them. Here, you have to sit down with three other players and—

KELLIE: Yeah, socialize.

HEATHER: You usually introduce each other. I think it's just a different thing altogether. It's a different scenario.[8]

▶ 4.3

Throughout the interview, Kellie and Heather emphasized the importance of the "band" factor in *Rock Band*, in contrast to the star-soloist karaoke model. Kellie told me, "If I play guitar, I'll go over to the drummer and kind of rock out—or if you have the bass and the guitar, you do the back-to-back thing and have fun, like you're really in a band." The game software lays the groundwork for band sociability by offering incentives for ensemble interaction; players are rewarded with crowd noise, bonus points, and on-screen "unison bonus" alerts when the whole group is playing well or when multiple band members deploy the high-scoring "overdrive" mode at the same moment. ("Overdrive" is *Rock Band*'s version of the "star power" feature in *Guitar Hero*; players gradually build up "overdrive" energy by playing well and can then choose when to deploy that energy.) Players activate "overdrive" mode by lifting the guitar neck, striking a cymbal at the end of a drum fill, or whooping into the microphone; band members often communicate verbally or with eye contact to coordinate overdrive deployment. When things are not going so well for the band—when one member is struggling and at risk of failing out of the song—a player who is doing better can deploy his or her overdrive energy to "save" the struggling bandmate and ensure that the group gets to complete the song. These features cultivate a sense of mutual obligation and collaborative effort among band members. As one player told me, "Getting completely caught up in the visceral 'feel' of a particularly gripping song and the energy being put into the experience by the others is very rewarding, even if it's a 'simulated' performance" (survey follow-up email correspondence).

All of my bar-night interviewees described *Rock Band* as an effective social icebreaker, noting that the game offered a focused activity and a sense of common purpose in place of awkward getting-to-know-you conversation. Brian told me that playing in a group created an affinity bond with other players, even if each group member was struggling to keep up with the streaming notation:

Even if you are focused on the music, I think just even standing up there with three other people doing the same sort of thing, you still have that common experience bond with other players. So whether or not you're actively engaging with your other players or not, you're still sharing something together; you're all creating something together, which, it does form that minor bond with strangers that you wouldn't normally have otherwise. . . . I think once you get comfortable with the game, it frees up your brain a little bit to really bond with the other players.

While karaoke certainly has its own social-bonding aspects (Mitsui and Hosokawa 1998, Drew 2001), *Rock Band*'s emphasis on ensemble performance rather than star turns makes it stand apart from its well-established schizophonic predecessor. The game nights brought out rock-star physicality in some performers, but it's worth remembering that apart from the occasional singer who turned to face the crowd in the rest of the bar, virtually all players had their backs to the audience—an audience that was only occasionally paying attention in any case. While playing in public still had the power to inspire some performance anxiety and adrenaline, bandmates were mostly playing for each other and themselves—to blow off steam, engage with a favorite song in a new way, or "hear new songs and discover new people," in Mike's words.

I returned to Our House East several times in the summer of 2009 and made multiple visits to two of Shaun and Mike's venues in other parts of town. On each occasion, I encountered a friendly, accessible musical scene with a tight-knit group of friends at its core—more like an open Irish session than most karaoke nights I've experienced, and with fewer barriers to entry than either of those other participatory bar events. Much to my surprise, the scene reminded me of the participatory tradition that was the focus of my first major research project: Sacred Harp singing, an American vernacular hymnody tradition that is open to anyone, regardless of perceived musical expertise, and that revolves around drop-in community "singings" rather than rehearsed performances for an audience (Miller 2008b). Given this tradition's strong DIY, anticommercial ethos (not to mention the religious song texts), many participants would find the comparison to a mass-produced rock-oriented video game bizarre. Yet at the *Rock Band* bar nights, I encountered a familiar sense of openness, community, and passion for collaborative music-making, where newcomers were urged to join in and assured that "anyone can do this."

A DAY WITH THE SCORE-ORIENTED: *ROCK BAND* TOURNAMENT PLAY

That same summer I experienced a very different version of *Rock Band* public performance by attending Game Unicon, a video-game convention held in an off-the-highway hotel deep in the Boston suburbs. I arrived around noon on a Friday; the three-day convention was just getting rolling. When I entered the crowded conference room devoted to *Guitar Hero* and *Rock Band*, I was immediately conscious of my age and gender. Women were very few and far between; a couple of moms came to watch their sons compete, but during my visit there were no female players. The vast majority of players were white teenage boys, some looking no older than 11. There were cash prizes in the offing, but apparently they weren't high enough to entice many adult players to take a day off from work.

I was at the convention to see the *Rock Band* individual instrument tournaments on guitar and drums. Five gaming stations were set up around the perimeter of the room, each occupied by two guitarists or two drummers competing to earn the highest score on a song. Although both members of each pair were playing exactly the same part, each player had his own on-screen notation track to follow (allowing him to make his own strategic choices about when to deploy the "overdrive" score-multiplier mode). The guitar finals were in progress on some stations; at others, players were warming up for the drum tournament. The room was loud, full of the clacking of fret buttons and strum bars and the dull thump of drumheads. The game music on each TV was turned down fairly low, so only the players directly in front of the speakers could hear it very well; otherwise it would have interfered with the experience of neighboring players. (This problem might have been solved by having competitors wear headphones, but then the audience at each station wouldn't have been able to hear the music at all.)

Players were required to bring their own instrument controllers; plastic guitars leaned against the walls, and a fleet of drum kits waited in a corner. Players with modified kick-drum pedals showed off their equipment and invited others to try out their kits, sometimes offering to sell their work after the convention. Some people had "premium" drum controllers (e.g., the Ion "Drum Rocker"), which were more responsive and much more expensive than the kit that comes with *Rock Band*. At first I was surprised that modified controllers and nonstandard drum kits were allowed, since they seemed to make for an uneven playing field. But of course some musicians have much better instruments than others, and athletes are allowed to compete in performance-enhancing shoes or swimsuits. Competitive *Rock Band* players devote a lot of time to mods, customization, and repairs (see Scorehero.com 2010). Tournament organizers were not willing to put convention-owned controllers through the sustained abuse of tournament play or to take responsibility for resulting technical snafus. They could not realistically require players to own a pristine, unmodified controller for competition purposes.

The basic rules had been posted on the convention website and several gaming forums in advance:

Rock Band Individual Instrument Tournaments Format
    Score Duel Double Elimination

You Must Play on Expert Difficulty

Each Match Is Best 2 out of 3 Songs

The Player on the top of the Bracket can choose to pick which side he is on or pick the first song. The first song is then played.

The Second Player then picks the second song.

In the event that each player wins a match, a third song will be chosen at random using Random.org; that song must be at least Moderate Tier in difficulty.

All Rock Band 1 and 2 Disc Songs will be available to pick. No DLC . . . there will be some there for freeplay but not in the Tournaments.

THE ROCK BAND FULL BAND TOURNAMENT WILL BE SEEDED BY THE GUITAR-ONLY TOURNAMENT ON FRIDAY!

(GameUniverse 2009)

The players were totally score-oriented, as one would expect at a tournament that didn't award style points. (At some tournaments, game points and judge-assessed style points are combined to arrive at a final score.) Competitors focused on the screen and never indulged in flamboyant rock-star moves, though I saw several people twirling their drumsticks or engaging in minimalist head-banging as they waited to play. Audience members with video cameras kept their focus mostly on the TV screens, not the players; they documented the steadily rising scores and the moments when something went awry for one of the competitors. (Numerous *Guitar Hero* and *Rock Band* videos on YouTube adhere to this format.) In one player-uploaded video from GameUnicon, the camera doesn't stray from the screen until the match is over, but the audience's excitement is audible; gasps, whoops, and applause gradually build in the closing bars of the song as a very close match reaches its nail-biting conclusion. One of the players posted a comment on the YouTube video: "I was flipping out inside xD I always end up choking on that part" (Brocky1213 2009; the emoticon "xD" indicates eyes squeezed shut and mouth open in laughter or excitement).

But while players didn't showboat like rock stars, the sheer physicality of their gameplay was striking. Waiting drummers clutched their sticks intently, standing behind active players and air-drumming along with the on-screen notation. Drummers had distinctive physical styles and approaches to drumhead rebound; some were choppy, heavy hitters, others fluid and graceful. No one was sight-reading; playing at this level requires practice, memorization, and strategic analysis of high-scoring song segments. (The "No DLC [downloadable content]" rule ensured that players couldn't use obscure repertoire choices to unseat their opponents.) Audience members were connoisseurs of the most difficult parts of each song; they tensed up in anticipation of tricky passages and shook their heads in appreciative awe when a player made them look easy. During the guitar tournament, people moved around to the side of the competitors to peer at their fretting hands, checking out their fingering techniques. Score-oriented players may look like glassy-eyed, interchangeable automatons to the uninitiated, but their approaches to competitive play are personally distinctive and adrenaline-fueled—they are "flipping out inside." As Keil wrote in an account of Japanese karaoke performances, "We are likely to notice and be concerned by the conformist aspects of this process, the adaptation to the machine, the preset accompaniment, the star model followed closely, but in its own context this singing can be interpreted as an assertion of individualism, skill, personal competence before others in a demanding situation" (1984:95).

Virtually all the competitors had chosen to play metal songs, which supply the most technically difficult guitar and drum parts in the *Rock Band* repertoire. Many players were also wearing black T-shirts advertising metal bands whose songs are featured in the games. Mastodon and Cannibal Corpse were popular options; Mastodon's "Colony of Birchmen" is part of the on-disc repertoire in *Rock Band 2*, while Cannibal Corpse's "Hammer Smashed Face" is available as downloadable content for an additional fee. In the age of digital discovery and media convergence, it's quite possible that these young players discovered the music through the games rather than first experiencing it as listeners—though of course the available musical repertoire also drives sales of the games (Jenkins 2006a, Jennings 2007). For a young tournament player in a Mastodon T-shirt, score-oriented *Rock Band* play probably represents only one of many portals to engagement with heavy metal. The first page of YouTube search results for the song "Colony of Birchmen" offers several more, including the band's official music video, live concert footage, and amateur cover versions (along with numerous *Rock Band* gameplay videos).

After a while, something about this cacophonous hotel conference room filled with fidgety teens playing plastic instruments felt eerily familiar to me. I realized that it reminded me of my All-State clarinet auditions in junior high and high school. There were no Cannibal Corpse T-shirts there—band and orchestra kids generally occupied a different niche in the teenage social ecology, and even the metalheads among us would have dressed a bit more formally for the audition judges. But there was the same charged atmosphere of performance anxiety, the same high-volume heterophony generated by compulsive repetition of tricky musical passages, the excitement of meeting peers who shared one's interests, and the daunting realization that one might not be the biggest fish in the pond.

I chatted with one of the proud moms at the *Rock Band* tournament as her son made it through several rounds of competition; he was visibly sweating from the exertion of playing drums, and he enthusiastically shook hands with his opponent at the end of each song. She mentioned that he also played several traditional instruments, including non–*Rock Band* drums, and was hoping to study at the Berklee College of Music someday—but it wasn't yet clear how serious he was about it. Meanwhile, playing these games was simply another aspect of her son's musical life, and she was happy to support it. She made me wonder: If that 14-year-old boy never played a real drum kit or went to Berklee, but put in hours and hours of disciplined practice on his plastic kit, gained an intimate knowledge of hundreds of songs across several genres, developed close friendships and respectful rivalries with fellow players, became a confident public performer, and expanded his multimedia skills by editing videos of his gameplay to post on YouTube, wouldn't his *Rock Band* expertise still be worth cultivating, according to the same evaluative norms that parents and teachers usually use to assess athletics, band, dance classes, private instrumental lessons, and other extracurricular pursuits?

My research is not about child development, nor about finding evidence to vindicate gaming as a wholesome, character-building experience. Because I did not have Institutional Review Board research-ethics approval to include minors in my research, I couldn't interview the tournament players. Still, my experience at Game Unicon shed some light on the role *Rock Band* played in the everyday social and musical lives of these teen boys. As they each worked to perform their score-oriented mastery of a specific catalog of songs, they were also cultivating relationships with each other, participating in collective consumption of the products of a complex multimedia industry, and developing their "sensational knowledge" (Hahn 2007) of heavy-metal masculinity, with its "conflicting paradigms regarding assertive, spectacular display and rigid self-control" (Walser 1993:108). Tournament players are "serious" gamers, but they are no less socially or musically engaged than "casual" players at *Rock Band* bar nights.

LIVENESS, SCHIZOPHONIC SPECTACLE, AND ROCK DRAG

Like album covers, videos, and live concert tours, *Guitar Hero* and *Rock Band* are now among the commercial products that tend to shore up rock music's status as a performing art, as opposed to a studio-recording art (Auslander 1999:65). The guiding narrative of the games reproduces the classic rock-career model of building a fan base through long tours and live performances—despite the games' own obvious reliance on sophisticated studio-recording technology. Players advance through each game by using "career mode," which consists of performances at increasingly large, prestigious, and geographically distant concert venues. The games not only provide engaged and vocal virtual crowds but are also designed to attract human spectators, who watch the antics of the on-screen avatar band. This spectacle is wasted on many players (because they need to stay focused on the streaming notation track until they have memorized a song), but it entertains friends, family, and coworkers—and once an audience has gathered, many players feel pressure to put on more of a show themselves. Following the lead of the games' designers, these players tend to integrate indices of "live" rock authenticity into their performances. As Simon Frith observes, "Rock performers are expected to revel in their own physicality. . . . Rock acts conceal not the physical but the technological sources of their sounds; rock audiences remain uneasy about musical instruments that appear to require no effort to be played" (1996:124–125; cf. Gay 1998).

At the same time, players are aware that *Guitar Hero* and *Rock Band* celebrate models of musical creativity, originality, and authenticity that cannot be realized in the context of their own game designs (see Chapter Three). Players develop the kind of virtuosic technical precision and sight-reading finesse that could garner limited praise in classical-music contexts, but they can never live up to the improvisational-genius model of the rock guitar hero. In subtle ways, these games continually remind players that what they're doing is *not* really playing rock music; the games often generate respectful appreciation of the gap between the player's performance and the recorded musician's performance. Player after player has hastened to assure me that he or she "understands the difference" between playing these games and being a real rock performer.

For rock-oriented players like Freddie Wong, that difference can manifest itself as a kind of rock drag. Like drag performances, rock-oriented schizophonic virtuosity is plainly discomfiting to some viewers; not only is the performing body almost entirely severed from the musical sound, but the physical performance gestures draw attention to their own theatricality. (This unsettling quality was prefigured in early electric guitar performances; see Waksman 1999:129.) The term "drag" seems particularly well suited to these performances because ideologies of gender and sexuality also play an important role here. As André Millard and Rebecca McSwain note, "the erect guitar" has long been "an essential part of the formalized ritual of the rock concert," contributing to the process by which "the meaning of the sound meshed with contemporary notions of masculinity" (2004:158). *Guitar Hero*'s focus on this instrument undoubtedly enhanced the game's appeal to the young, male players who weren't interested in games like *Karaoke Revolution*.

But for many *Guitar Hero* commentators, there seems to be a transparent connection between playing a real guitar and being a real man. A fake guitar implies a false masculinity, and the "star power" tilt creates an offensively fake erection. I have read countless homophobic and feminizing insults in YouTube comments about these performances. For instance, one commenter on Freddie Wong's video wrote, "yur a gay ass mother fuckin fag ill bett you 20 mother fuckin dollars that this shit is fake" (GHman64, comment posted on freddiew 2006).[9] Homophobic insults are a common feature of YouTube comment threads, but the connection hardly seems incidental in this case. Sexuality-oriented assessments also appear in media reviews. *San Francisco Chronicle* reviewer Peter Hartlaub wrote, "Playing a Guitar Hero or Rock Band guitar is a fairly effective form of birth control. Seriously, look at yourself in the mirror. No one who sees you playing this thing will want to have sex with you" (2007). Most famously (at least for fans of American television), an episode of the animated sitcom *South Park* was titled "Guitar Queer-o." When the show's characters earn 1 million points playing a *Guitar Hero* clone, they are rewarded with the on-screen text "Congratulations! You are fags!" (South Park Digital Studios 2007).

Steve Waksman and Leslie Gay have described the naturalization and incorporation of the electric guitar, which allow the instrument "to invest the body of the performer with meaning, to confer upon it a unique identity whose authentic, natural appearance works to conceal its reliance upon artifice and technology" so that "every note takes explicit shape as a physical manifestation of the performing musician" (Waksman 1999:5, 243). Rock guitarists "demonstrate authority over the technology by making it less an extension to one's body than part of the body itself" (Gay 1998:85). The *Guitar Hero* controller undoes this incorporation, creating a marked separation between embodied performance and musical sounds. The resemblance to the destabilizing effects of cross-gender performance is striking; as Judith Butler writes, drag denaturalizes sex and gender "by means of a performance which avows their distinctness and dramatizes the cultural mechanism of their fabricated unity. . . . Although the gender meanings taken up in these parodic styles are clearly part of hegemonic, misogynistic culture, they are nevertheless denaturalized and mobilized through their parodic recontextualization" (1990:138).

In some sense, then, homophobic critics are right to suspect that there's something queer about *Guitar Hero* and *Rock Band*.[10] The games both cite and encourage camp-inflected performances, in which "hegemony is queered, denaturalized, and, thus, subverted through overarticulation" (Devitt 2006:33, following Niles 2004 and Babuscio 1984). According to Moe Meyer, the queer signifying practices that constitute "camp" pose an ontological critique by displacing "notions of the Self as unique, abiding, and continuous while substituting instead a concept of the Self as performative, improvisational, discontinuous, and processually constituted by repetitive and stylized acts" (1994:3). In general, video games and other forms of interactive digital media tend to draw attention to the unstable, performative nature of identity, through role-playing and character-customization features (see Chapter One). In *Rock Band*, players not only choose the gender, body type, clothes, and instruments for their avatars but also must select a physical performance style, choosing from rock, punk, metal, and goth "attitudes" that govern the avatar's physical mannerisms, stance, and affect. Of course, the movement practices of video-game avatars always draw on "a more general set of reflected and represented cultural choreographic practices" (Burrill 2006:19). However, *Guitar Hero* and *Rock Band* present a special case because the player's performance may closely mirror or even outdo the avatar's. One may integrate "goth attitude" into one's own physical performance on guitar or drums, rather than simply directing an avatar's movements through button-mashing on a standard game controller. The games invite players to make a spectacle of themselves, something Robert Walser identifies as at odds with "a patriarchal order that is invested in the stability of signs" (1993:108). Both naturalized rock authenticity and its associated masculinist gender ideologies are potentially at stake here, as in the glam rock performances that the games commemorate.

Considered in these terms, schizophonic performance in *Guitar Hero* and *Rock Band* has the markings of "the camp trace" or "residual camp"—Meyer's terms for un-queer appropriations of queer praxis. Meyer's account of the queering effects of "residual camp" helps explain the homophobic responses to spectacular game performances. Following Andrew Ross, he argues that when the un-queer camp cognoscente appropriates queer signifying practices, "because the queer constitutes him-/herself processually, the un-queer is now unwittingly performing the queer. The final effect is the reproduction of the queer's aura by the un-queer camp liberator who has been transformed into a drag queen with no other choice but to lip-synch the discourse of the Other" (1994:17, following Ross 1989:Ch. 5). Or, to return to *South Park*, "Congratulations, you are fags!"

My aim here is not to valorize these games and their players by identifying them as inherently subversive or resistant, in the manner of early cultural studies scholarship. As Butler writes of camp, "Parody by itself is not subversive, and there must be a way to understand what makes certain kinds of parodic repetitions effectively disruptive, truly troubling, and which repetitions become domesticated and recirculated as instruments of cultural hegemony" (1990:139). Freddie Wong and his ilk are savvy parodists of rock authenticity, but one could also argue that their performances "take rock's ideology of authenticity as [their] point of reference and [are] therefore allied with that ideology" (Auslander 1999:101). It's also important to note that some players adopt quite sincere and serious approaches to the game content, especially in terms of respect for the music; many others readily switch between sincere and ironic stances, depending on the particular song or performance context (like the *Grand Theft Auto* players discussed in earlier chapters). The point is that these games give players such rich scores and scripts to interpret, with respect to ideology as well as musicality and theatricality. Rather than creating musical automatons, *Guitar Hero* and *Rock Band* have inspired players to craft a tremendous range of performance styles that in turn garner widely divergent audience responses.

VIRTUAL VIRTUOSITY/VIRTUOSITY WITHOUT VIRTUE

Freddie Wong has had ample feedback from online audiences. His "YYZ" video generated over 43,000 comments between fall 2006 and spring 2011. Consider a few comments posted during the summer of 2010:

bpiech1: Too bad that your birth was the first and last time that you will ever see female genitalia.

cozanger21: In the time you spent getting that "good," then thinking up, creating, and posting this video, you could probably have cured breast cancer. You are not awesome. You are pathetic.

07Attila94: more proof that your human is that you missed at 2.17—but thats still pretty ace, even so i think you seriously need to look at the time you spend on your hobbies . . .

Here we see an emasculating taunt, a reprimand that charges Freddie with wasting time that he could have spent on a socially valuable endeavor, and finally an ambivalent assessment: The last commenter seems to conclude that Freddie is just good enough at *Guitar Hero* to be criticized for spending too much time on it. Note that this commenter draws attention to a technical flaw in Freddie's playing and describes it as "proof that [he is] human." This is a backhanded compliment; it both undercuts Freddie's claims to rock god status and offers confirmation that the high score wasn't faked. Freddie and his friends anticipated this response in creating the video:

When we made the video originally, we actually planned for "talking points" or things that we hoped people would point out, in an effort to generate conversation. . . . One thing that worked very well was that I actually missed notes on purpose—had I played the song perfectly, people would immediately start to think it was faked somehow (the debate in online videos of veracity being something that comes with the territory of apparently amateur user-generated content). . . . At the time the video came out, it was fairly well known there was a hack to have the game play the song for you 100% [accurately], so we wished to avoid that dismissal, which would cause viewers to not pass along the video (which is instrumental in popularity). As a result, every time someone calls the video fake, another user replies that the fact that I missed notes means it wasn't fake, so our strategy worked out perfectly. (Miller 2007a)

Freddie divided the thousands of comments on this video into several categories, including calling him names, defending him, bragging about the commenter's own skills, and telling him he should play a real guitar instead. In fact, he has been playing electric guitar far longer than *Guitar Hero,* and he eventually added a link from his "Rush YYZ on Expert" YouTube video to one that shows him rocking out on a Fender (freddiew 2008b; see Figure 4.6, ▶ 4.4). Early in this video, which is titled "True Guitar Heroism," Freddie offers an irony-drenched, doth-protest-too-much defense of his heterosexual prowess, exhorting viewers to click on a link that will provide "video proof . . . that I hang out with hot chicks all day, all the time." The guitar performance that follows is a campy extravaganza, laden with as many rock clichés and fetishistic fretboard close-ups as the "YYZ" production. At the end, Freddie looks straight into the camera and asks, "Who's a faggot now?"

Figure 4.6
Still from Freddie Wong's "True Guitar Heroism" video (freddiew 2008b). (*Screenshot by the author*)

By way of contrast, consider Chris Chike, a teenager from Minnesota who has been recognized by Guinness World Records for his high score on the Dragonforce song "Through the Fire and Flames" in *Guitar Hero III* (Guinness World Records Gamer's Edition 2008). He uploaded a YouTube video documenting his first 100 percent accurate performance of this song in June of 2008 (iamchris-4life 2008; see Figure 4.7, ▶ 4.5). In the video, the camera is trained on the TV screen in Chike's living room; Chike only appears from behind and at the margin of the screen, where the fingers of his left hand move over fret buttons at an extraordinary speed. By spring 2011, this video had been viewed over 5 million times and had generated over 73,000 comments. Here is a sampling:

HaniHM10: tHtS so Inhuman xD

pieguy121: WITCH WIIITCH!!!!!!!!!!!!!!!

sulkyns: This is not fake, he is the best.

MrSouthEcko: ok its either a bot or ur satan. . . . lol jk nice vid tho

XXHystericAnarchyXX: Its a Bot, the parts he apparently acctivated star power were not hammer-ons and the only way he could hve activated Starpower [with]

Figure 4.7
Still from Chris Chike's "Through the Fire and Flames" video (iamchris4life 2008). (*Screenshot by the author*)

out lifting up [the] guitar (As he did) Would be to hit [the] select button and he would miss a note.

xXDarkNinjaKorakuXx: Honestly dude, lots of other people have done this too. It should be to the point where if someone else does this, they aren't panting over the fact they accomplished a common goal that most no-lifers like yourself achieve. Try doing something real with your life.

slashguy105 (reply): damn. you must NOT like this game huh? id like to see you achieve that "common goal." until you do that and post it on youtube so we can see, you have no room to insult him. dont be harsh cause hes good at a game. people who go around trash talking peoples achievements because they cant do as well as others at something are the people without lives. go hate on someone who actually sucks. video games exist. that makes them real. you have no idea how hard this is until u try.

Notice how the discourse revolves around Chike's purported inhumanity, supernatural powers, and accusations that he is using a bot (a computer program that automatically hits the right notes). Using more sophisticated language, a *New York Times* reporter touched on similar themes in a profile of Chike:

The sounds were coming from the basement: a rapid, staccato clicking of plastic against plastic, hundreds of times a minute, too quick and orderly to be described as a rattle and too rhythmic to be considered noise.

In the downstairs den of his family's home, Chris Chike was sitting in an easy chair with a toy guitar across his lap. While his eyes were trained on a big-screen television inches away, his hands were frantically working the bruised plastic instrument held together with masking tape. His left hand was tapping manically at five colored buttons on the guitar's neck, while his right hand made graceful leaps between the neck and a large black button on the guitar's body, where a real musician would be strumming at strings. His movements were precisely choreographed to the action on his television set, where color-coded musical notes stampeded down the neck of a simulated guitar in time to a relentlessly bombastic heavy-metal ballad. (Itzkoff 2008)

This account captures the tone of many media assessments of *Guitar Hero*. Itzkoff employs ominous language—unclassifiable mechanistic sounds emerging from a basement, a "bruised" guitar—and focuses on a "frantic," "manic" player who is not a "real musician" but an expert automaton. Later in the piece Itzkoff acknowledges Chris Chike's "preternatural dexterity" and "peculiar blend of athleticism and showmanship," but he repeatedly differentiates Chike's performance from musicianship: "He is clearly not making music with the instrument, but his performance is a feat of coordination in its own right." The game code choreographs Chike's movements. What could be further from the spontaneity, originality, total musical engagement, and flamboyant iconoclasm of a real guitar hero?

Freddie Wong and Chris Chike represent two distinct models of embodied virtuosity employed by *Guitar Hero* players and their audiences. Wong's spectacular performances draw on the rockist tropes of heroic musical genius and radical individualism that are in turn indebted to the hero discourse of nineteenth-century Western European musical virtuosity. There's no arguing with his visceral energy, yet Wong's virtuosity is always already undermined because it's patently inauthentic, untrue to its own Romantic ideals. It is rock drag. By contrast, Chris Chike's score-oriented virtuosity has a cyborg quality. He's merging with the game machinery (cf. Gay 1998, Svec 2008). Chike's virtuosity is unspectacular almost to the point of invisibility. His accomplishments are quantitatively verifiable by means of his game scores, yet critics constantly accuse him of faking those scores with technical trickery—and those who don't question his scores chastise him for wasting his time and talent.

Whether they are score-oriented or rock-oriented, the more accomplished players appear to be, the more they risk disparagement. Even fans of the games often back away from defending high-level players. For instance, those who think the games are great for people who are "not musical" are less comfortable with *Guitar Hero* virtuosity; they often suggest that to play at this level, one must in fact have some natural musical talent, which shouldn't be squandered on video games (cf. Kingsbury 1988). As Shaun Scovil told me, "I mean anybody who says 'I'm great at this,' everybody else is gonna come back and say,

'Well, you got so much free time on your hands.' There's no gratification. Even if you win, you kind of lose. People are like, 'Oh great, you're really good at the fake music video game.'"[11]

But performers like Freddie Wong and Chris Chike are also cultivating audiences who are connoisseurs of alternative forms of virtuosity: virtuosic rock drag, cyborg virtuosity, and other forms of collaborative, dispersed, and mediated performance. In 2008, Freddie participated in an MTV-sponsored televised *Rock Band* tournament. (Because MTV owns Harmonix Music Systems, the tournament functioned as an extended advertisement for the game.) The YouTube video of the winning performance, in which Freddie's band Hellanor Brozevelt performed the Ramones classic "Blitzkrieg Bop," provoked several exchanges like this one:

almostwilt: these guys fucking suck

bigboee415 (reply): keep in mind theyre doing it live bitch[12]

Another defender addressed the score-oriented/rock-oriented divide: "To all the haters: This video isn't about score. It's about performing like rock stars. They had the crowd going nuts when they were using PLASTIC INSTRUMENTS. That's way more impressive than note streaks and FC's."[13]

Performances like these perfectly illustrate Weber's observation that "a major function of the theatrical in an age of electronic media is to articulate the ways in which sites—and sights, but also sounds and other 'sensations'— remain linked . . . to bodies, although not necessarily to *human* bodies as traditionally understood" (Weber 2004:48). Hellanor Brozevelt's spectacular theatricality made it possible for them to forge a compelling connection between a prerecorded song and their live, performing bodies, without denying or concealing the technological mediation involved. Built from the stylized gestural repertoire of rock performance and lubricated with genuine sweat, their performance drove the crowd wild.

CONCLUSION: SCHIZOPHONIC PERFORMANCE AND RETHINKING THE REAL

A major question in the reception discourses for *Guitar Hero* and *Rock Band* has been whether the games might revitalize and perpetuate "real" rock performance and fandom. For every media story that wonders why players don't pick up a real instrument, there is one that celebrates the players who have— like the *Guitar Player* writer who asserted, "[*Guitar Hero*] itself is a bit of a hero, as it leads generations of game-console fiends to consider the joys of actually playing the real thing" (Ross 2008:63). Sleater-Kinney guitarist Carrie Brownstein also adopted this position: "With so much of music blurring the lines between ersatz and authenticity, at least the Rock Band game is a tribute to rock, rather than an affront. . . . Maybe by pretending to be in a band, there will be those who'll find the nerve to go beyond the game, and to take the brave leaps required to create something real" (2007). Capitalizing on that hope, the website GuitarHeroTab.com provides real guitar tablature for songs featured in *Guitar Hero* and *Rock Band;* the site promises, "If you want to be a *real* Guitar Hero or play in a *real* Rock Band, we've got all the tunes you need to get rocking on stage!" (2008). Other companies have worked to develop guitar controllers with complex fretting systems and strings, which can be plugged into a game console or an amplifier (e.g., Inspired Instruments 2010). As I was finishing the manuscript for this book, Harmonix had just released a similar controller for use in *Rock Band 3*'s "Pro" mode, which the company promised would "empower players to develop actual musical skills" (Harmonix Music Systems 2010b). These hopeful stories and crossover-oriented business models offer an eerie echo of early-twentieth-century discourse about the phonograph's potential to make all Americans more musical— although phonograph advocates had a different kind of "good music" in mind (Katz 2004:Ch. 2).

Writing about earlier representational technologies, Michael Veal makes observations that might shed some light on "realness" and simulation in the *Guitar Hero* and *Rock Band* context: "Virtual technologies such as sound recording and film were often misunderstood in their early years as serving purely documentary functions; their creations were often dismissed as inferior simulations of reality. A more expansive take is that creative manipulations of these technologies in fact create new forms of reality (that is, new ways of 'hearing' the world) within which they function as 'prosthetic' devices, ultimately extending human sensory perceptions into new areas" (2007:218; cf. Murray 1997:103). *Guitar Hero, Rock Band*, and similar video games are working in exactly this way to create new modes of musicality. Playing *Guitar Hero* and *Rock Band* is deeply theatrical, yet there is no simulation or false promise in the games' offer of musical collaboration, and no lack of felt "reality" in the analytical, emotional, social, and physical engagement with music that they engender. Player Chris Sanders put it more elegantly: "The energy and commitment to the music is quite real, even if the instruments are not" (survey follow-up email correspondence).

Playing *Guitar Hero* and *Rock Band* isn't just like playing a real instrument, but it's nothing at all like just listening to music. The affective experience of making music is bound up with embodied performance, and these games compel bodily engagement. As Anna Munster suggests in her book *Materializing New Media*, when we consider the relationship between body and machine, we need to acknowledge that embodiment is a dynamic property—"not a property that 'I' have but something I produce, that is produced in relation to other bodies and machines" (2006:149). The underlying idea here has been a basic assumption in several disciplines for decades, and it rings true at a visceral level for those of us who play musical instruments: We know how machines can reshape our bodies. But some forms of embodied experience are certainly still considered more authentic than others, especially when it comes to assessing musicality. As reactions to *Guitar Hero* and *Rock Band* demonstrate, liveness still matters, and schizophonic performance can make people very uncomfortable.

So why is that? Playing around in the grey area between the virtual and the visceral is an everyday experience for anyone with access to interactive digital media, yet *Guitar Hero* and *Rock Band* remain provocative. Why have these games provoked heated debate, whereas the sports games for the Nintendo Wii are praised for getting gamers out of their chairs? Perhaps it's because nobody imagines that it would be possible to mistake Wii tennis for real tennis; after all, there's not even a real ball or a tennis racket. But *Guitar Hero* and *Rock Band* put instruments in players' hands, and when they play, undeniably real music comes out of the speakers. The games also have provocative power because they revolve around rock performance, a genre deeply rooted in discourses of radical individualism, heroic creativity, and embodied liveness. *Guitar Hero* and *Rock Band* are encouraging millions of players to celebrate and commemorate these core rock principles—the games are *about* guitar heroism—yet the nature of the gameplay experience tends to undermine those principles. The games offer vivid and visceral demonstrations that a player's embodied experience of musical performance can be a collaborative endeavor, rather than a heroic individual achievement, and that one's collaborators can be widely dispersed in time and space. As players and their audiences engage in schizophonic performance, they collaborate with game designers and recorded musicians in mending the schizophonic gap, stitching recorded musical sound and performing bodies back together. The satisfactions of this endeavor are not necessarily undermined by the fact that the player doesn't occupy the same performing body as the person who first produced the music.

A rock orientation has proved particularly effective in generating meaningful musical experiences for players because the games draw on such broad and deep reservoirs of existing musical and cultural knowledge. As designer Rob Kay freely acknowledged, rock's well-defined ideological profile did a great deal of work for the game designers, because so many people instantly know "how to get into it." Players' own musical histories and their engrained understanding of rock performance inform their every move. Some players still judge their own game experiences according to familiar rock ideologies of originality, creativity, and heroism. But many have concluded that these authenticity discourses, like rock music itself, constitute a performance repertoire rather than the rules of the game.

PART THREE

Playing Along with Communities of Practice

CHAPTER 5
Music Lessons 2.0

"Okay! As we continue our guitar journey, we need to talk about how you're going to be attacking the strings. And I'm going to recommend that you use a pick." David's tone is upbeat and encouraging, as always, and he seems to be looking right at me—his ability to make eye contact with the camera is uncanny. Propped on his right thigh, his acoustic guitar looks like a natural extension of his body.

I have a borrowed guitar on my own lap, and I've set out a few picks on my desk. Though I've watched hundreds of guitarists perform over the years—and I have David's example right in front of me on the computer screen—I'm having a hard time figuring out even how to hold the instrument. In the previous lesson, "How to hold the guitar," David emphasized that I should hold it close to my body so that it doesn't "dance around," and also that I shouldn't hunch or lean over the guitar. But these instructions seem somehow incompatible with my (female) anatomy. Nevertheless, I've moved on to "How to hold the pick."

"How to hold the pick," Lesson 11, is almost 11 minutes long (▶ 5.1). As I click the "Play" button, I wonder how it could possibly take someone 11 minutes to explain how to hold a guitar pick. But as the video runs and I follow David's directions on getting to know the pick, I realize that he's compensating for the restrictions of this virtual lesson interface. We're each holding a genuine three-dimensional guitar and pick, but he can't reach through the screen to adjust my fingers—and since the lesson is part of a prerecorded curriculum, he can't see my fumbling or offer specific feedback. When the lesson ends, I hit "Play" again and take notes:

> Early on he really focused on the tactile nature of the pick—especially the fact that it might be slippery, and therefore picks with raised letters might be an advantage at first. He asked the cameraman to zoom in on the pick and demonstrated subtle

differences in the angle of the tip. He showed that you should hold the pick tight enough that you can't pull it out with your other hand, but not super-tight because it's important to stay "natural." This reminded me of two things—how jewelry sales clerks explain how a ring should fit (you shouldn't be able to pull it right off), and those exercises where you close your eyes and feel/describe all the sensory qualities of a raisin (except there's no eating the pick at the end). By zooming in on the pick and talking about texture, slipperiness, tension of the grip, etc., David really encouraged a lot of physical awareness and sensitivity to subtle differences in picks or pick-related technique. I wonder if a face-to-face teacher would just physically correct the student instead, or hold his/her hand up to the student's hand. (fieldnotes, December 9, 2009)

My previous guitar-playing knowledge was derived entirely from *Guitar Hero*. As noted in earlier chapters, some technical basics do cross over. For me, these included the distinct roles played by fretting hand and strumming hand. My main instruments are piano and clarinet, so when I first started playing *Guitar Hero*, positioning the fretting hand in advance of hitting the strum bar felt counterintuitive. However, since a *Guitar Hero* controller has no strings and the strum bar isn't detachable, the games didn't give me any advance training on holding a pick. As I followed David's directions and (in subsequent lessons) tried some vigorous strumming, it was surprisingly difficult to hang onto the thing. Paying attention to texture made a big difference: The raised letters on one pick gave me crucial tactile feedback, both for judging the right amount of tension between my fingertips and for feeling when the pick was sliding around in my hand. I can't say how or whether David's Lesson 11 was different from the way an in-person lesson might unfold, but I suspect that a typical private instructor wouldn't spend 11 minutes encouraging me just to explore what the pick felt like in my hand. And if I were paying by the hour, I might feel cheated if she or he did. I'd certainly feel more anxious about having my pick go flying across the room.

I began to explore online guitar lessons as an extension of my research on *Guitar Hero* and *Rock Band*. Initially, I did a YouTube search to find out more about "hammer-ons," which I first encountered as passages of miniature notes in *Guitar Hero* notation; during hammer-on sections, you can play by just hitting the fret buttons without strumming. This felt like cheating until I learned that it's also a standard guitar technique. YouTube taught me about hammer-ons, pull-offs, and "tapping," presenting a wealth of how-to lessons as well as videos of famous performers, such as tapping virtuoso Eddie Van Halen. Mixed into these same search results I also found lessons for conquering tough *Guitar Hero* solos, including fingering-pattern instructions for the hellish hammer-on introduction to "Through the Fire and Flames" (see Chapter Four). It struck me that taking video music lessons and playing video games might have a lot in common. I began to wonder about the *Guitar Hero* players who have taken up the challenge to "lrn2reeltar": Surely many of them would start this process on YouTube.

YouTube now hosts a vast selection of music-lesson videos. In July 2010, a search for "guitar lessons" yielded about 527,000 results. However, many of the most popular YouTube lessons and channels also function as teasers and community-outreach updates for other websites. The proliferation of these YouTube-linked sites exemplifies what Burgess and Green call the "permeability" of YouTube—the way it "connects with surrounding social and cultural networks, and users embedded within these networks move their content and their identities back and forth between multiple sites" (2009:66). Although running a stand-alone website burdens teachers with web-design tasks and procuring server space, it gives them much more control over content; they can offer supplementary materials that are not supported by the YouTube channel format, including downloadable sheet music, tip sheets, and message-board-style community forums. Running their own websites also allows teachers to earn more money for their work, by accepting advertising, charging students a subscription fee, or selling material goods through an online store-front. (I say "more" money because many YouTube users already earn a financial return by signing up to have YouTube display targeted advertising with their videos.) Moreover, by hosting their videos on their own servers, teachers can protect themselves from peremptory deletion of a YouTube account in the event of a copyright violation claim—a major risk for anyone who posts lessons that teach students to play copyrighted popular songs.

This chapter focuses on two music-lesson websites, NextLevelGuitar.com and OnlineDrummer.com, each of which has attracted thousands of students. I begin this part of the book with guitar- and drum-lesson websites because they offer productive continuities with the *Guitar Hero* and *Rock Band* chapters in terms of musical repertoire, instrumental technique, and player communities. *Guitar Hero*, Next Level Guitar, and OnlineDrummer all came on the scene around the same time (2004–2006), and many students sought out the lesson websites after first experiencing their instruments and potential performance repertoire through the games. David Taub and his business partner, Tim Gilberg, addressed this connection in a YouTube video in which they answered frequently asked questions from students; apparently people often asked them what they thought of *Guitar Hero*.

TIM GILBERG: It's all good! If you're involved in music, it's great. I started playing guitar because my daughter got Guitar Hero, and I was like, "If I can play, if I can practice Guitar Hero for an hour a day, then I can practice guitar." We're members of NAMM [the National Association of Music Merchants] . . . and they're wholeheartedly embracing Guitar Hero as a stepping stone, for not just kids but anyone. . . . If it gives you the stepping stone to move on to the guitar, great! If you just stay in Guitar Hero, great! . . . You're still doing music, you're still doing the beat, you're still doing rhythm, you gotta do it in time—

DAVID TAUB: It definitely helps, and it works the finger dexterity. But after a while, if you're playing Guitar Hero for an hour, two hours a day, you might as well learn guitar. (rockongoodpeople 2010; see ▶ 5.2)

The Next Level Guitar and OnlineDrummer sites illustrate how music teachers with minimal digital-media experience and financial capital are approaching some of the same challenges as major game-development companies like Harmonix Music Systems and Rockstar Games: how to create interactive content that is compelling enough to attract and retain students/players/customers, how to use social networking sites and media convergence to further experiential and entrepreneurial goals, how to negotiate the complexities of copyright law, and how to bridge the gap between the virtual and the visceral. After using these examples to explore some key affordances and challenges of online instrument lessons, in Chapter Six I will present three contrasting case studies of online pedagogy, focusing on "amateur-to-amateur" transmission.

PICK UP A REAL GUITAR: DAVID TAUB AND NEXT LEVEL GUITAR

On the day I first went looking for guitar lessons on YouTube, videos by "rockongoodpeople" were among those that rose to the top, through the complex algorithm of number of views, canny metadata choices, and (sometimes) paid placement that drives search results. By the end of 2010, rockongoodpeople's profile boasted almost 72 million total upload views—numbers impressive enough to keep attracting yet more viewers, in the virtuous cycle of YouTube celebrity. David Taub, the friendly guy featured in most of the "rockongoodpeople" videos, is a longtime guitar teacher who now works full-time producing content for his online video curriculum and song-lesson DVDs. (See Figure 5.1.)

David Taub grew up in New Jersey and began playing guitar in the sixth grade. His public elementary school had a few acoustic guitars and would loan them out for short periods to students "who the teachers identified as having a good week behaviorally."[1] (David noted that he was generally not first or second in line for this privilege.) He was "self-taught" for many years—that is, like many rock guitarists, he learned from recordings, casual interaction with friends, and feeling his way around on his instrument. Later, he went through a series of private teachers "before I could really find one that took it seriously enough and really put the time into the lesson planning and the curriculum, not just coming up with stuff on the fly to teach you." This early experience informed David's teaching approach. Throughout our conversations he kept coming back to the importance of careful, detailed, individualized lesson planning and the teacher's responsibility to the student: "So what a lot of people don't realize is that *you're paying them*. They're working for you, so you have to make sure obviously that

Figure 5.1
Still from a David Taub guitar lesson on YouTube (rockongoodpeople 2010). (*Screenshot by the author*)

you get what you want out of the lessons. . . . It's a two-way street." Before founding Next Level Guitar, David had been supporting himself with private teaching and performance gigs; for a while he spent two days a week driving around the San Diego area to teach students who were willing to pay for house calls. Eventually he built up a teaching reputation that allowed him to stop the travel-teaching and require students to come to his studio. But he could still only teach about 30 students a week, "because the lesson planning takes so long."

David spoke freely about "mediocrity in the teaching field" and the resulting high level of attrition among beginning guitar students: "You know, there's tons and tons of really great guitar players. . . . But teaching's an art, and I've been refining that art my whole life." Because he believed passionately in the efficacy of his own teaching method, he wanted to find a way to reach more people. One of his students was a professional freelance videographer, and David began talking with him about making some guitar-lesson DVDs; they shot several hours of film toward that project. In 2006, another student asked if it would be okay to film part of a lesson and post it on Google's video site. David agreed, and he was amazed when "it got thousands and thousands of hits." The student, Tim Gilberg, had a background in online marketing; he suggested that he and David work together on a guitar-lesson website. As David told me, "It's kind of a unique situation, where it was two lifetimes of work that we were able to put together to make this business happen—my lifetime of music knowledge and teaching and playing and ability to teach students in different ways to get them to learn and do it really well, and his lifetime of sales, computers, and marketing." They began working on a video curriculum, occasionally posting free videos on YouTube to build a following. Once they had made about 250 lesson videos for Next Level Guitar, they felt they had enough material to publicize the website and start charging for subscriptions. (David told me that he opted for the subscription model because he felt banner advertising would be a distraction from the lessons.) Various subscription options are available; I chose to pay $75 for three months of unlimited access to the video curriculum.

David and Tim started posting their videos on YouTube just when it was becoming a dominant online video site. Views beget views on YouTube. As more and more people started searching for instructional videos, and as media coverage of YouTube ramped up, their channel became extremely popular. To increase their audience, they experimented with the esoteric art of keywords and tags: "That's a whole art in itself, getting your videos to come up high in the rankings and pages. How many pages do you go through when you do a search—one, two?" Only six weeks after opening NextLevelGuitar.com to subscribers, David and Tim had hundreds of paying students signed up. By early 2010, at any given time there might be 600 students logged in. David told me that students come from the United States and Canada, the United Kingdom, Germany, Australia, Egypt, and "all over the world"—though the fact that the lessons are in English and are promoted on English-dominant websites has had a predictable limiting/filtering effect. Over 13,000 people have registered as users of the site's messageboard-style community forum, which is free. Paying subscribers also have access to a premium-content section where they can ask David questions. It is likely that many more people have explored the message threads on the forum without registering as users.

"THE PLACE TO STICK IT ONLINE": NATE BROWN AND ONLINEDRUMMER.COM

At the beginning of the lesson, Nate Brown is turned away from the camera. Wearing large headphones, he plays a samba beat on his drum kit. After about 10 seconds of repeating the beat, he swivels around to talk to me, but not quite all the way—it seems like he doesn't want to turn his back on his drums. Now I can see that his headphones are part of a headset microphone kit, which allows him to balance the volume of his voice with the loud drums and to play along with backing tracks when teaching song lessons. The headset microphone also imparts a formal, professional, recording-studio feeling to the lessons—or perhaps a televised-fitness-class feeling, depending on your point of reference. Like a step-aerobics or dance teacher, Nate offers a quick demonstration, then a slowed-down version; then he counts out the beat and off we go together.

As Nate explains the beat to me, he almost has to look over his shoulder to connect with the camera. For the first time I consciously realize that drummers work with their kits face to face, whereas a guitarist presents both her own face and her instrument's face to the audience. Nate wants students to have a clear view of his drumheads, cymbals, and sticks, so he has to sacrifice the intimacy of locking eyes with students through the camera—or he could use multiple camera angles and frequent cuts, but that would be both time-consuming to edit and potentially distracting to students. When I play along with David, I watch his face as much as his playing, trying to mirror his confidence and comfort with the guitar. When I play along with Nate, I'm mostly watching him from behind or to one side, so I try to inhabit his gestures and see through his eyes. (fieldnotes, July 19, 2010; OnlineDrummer.com 2010d, onlinedrummer 2010b; see Figure 5.2 and ▶ 5.3)

When Nate Brown was in elementary school, he spotted a neighbor engaging in a venerable form of virtual performance: "There was a gentleman that lived behind me. He had a drum set in his basement. . . . I'd sneak over and look in the window and watch him play. He'd play along with the radio, and that really interested me."[2] At the time Nate was taking organ lessons: "It's what my [divorced] mom could afford, since we already had the organ." But by the fifth grade he had persuaded his father to buy him a snare drum, and a couple of years later his mother and grandmother scraped together the money for a drum set. He took private drum lessons until he was about 16, played in several bands, and soon began teaching students of his own at a local music store. Nate loved performing, but his life circumstances made him cautious about the risks of that career path: "When I was a young age I had a daughter—which I'm happy for—and I decided

Figure 5.2
Still from a Nate Brown drum lesson (onlinedrummer 2010a). (*Screenshot by the author*)

maybe playing in a band and doing all that touring wouldn't be the best thing to do." He decided to pursue a career in education instead. Originally he wanted to focus on music education, but this track didn't seem much more secure than being a touring musician: "I realized that I have a daughter and I need a job as soon as possible. . . . You've got one [music] director per school or maybe even only one or two for the whole district, so it's going to be difficult to get a teaching position right out of school. So I switched to math and science, and now I'm a science teacher." As he worked toward his education degree, he continued teaching private drum lessons to help support himself and his family.

In college, Nate learned "a lot about teaching, but not only teaching—a lot about creating websites and designing online material." It occurred to him that he could use these skills to create a website for drum teachers and students. With some help from his brother (a professional programmer), he created OnlineDrummer.com in 2004, when he was 23 years old. He posted a new drumbeat to the site every week, including staff notation, tab notation, audio and video examples at two tempos, a genre designation, and a difficulty rating. (See Figure 5.3.) Before long, "Somebody emailed and said, 'Hey, can you teach me how to play [the AC/DC song] "Back in Black"?,' because he saw us putting up drumbeats each week. I said, 'Well, I'll put together a little lesson,' and I did, and put it online, and it was a hit." Nate posted the "Back in Black" video tutorial on both OnlineDrummer and YouTube; it's the oldest video on his You-Tube channel, posted on February 10, 2006 (onlinedrummer 2006; see ⏵ 5.4). By the end of 2010 this video had been viewed almost 300,000 times and the YouTube channel had over 13,000 individual subscribers.

As with David Taub's "rockongoodpeople" channel, the main purpose of the "onlinedrummer" YouTube channel is to drive more traffic to OnlineDrum-mer.com. All of the content on the OnlineDrummer site is freely available; Nate Brown is not using YouTube to attract paying subscribers. However, attracting page views is crucial, since the site is supported by advertising rev-enue. Traffic to OnlineDrummer in turn drives traffic to a newer site, Drum-Music.tv, where Nate sells lengthy video tutorials for individual songs

Drum Beat: "Bellin" (12-13- 2004)   Difficulty = 3/5
<u>Category</u>: | *Hip Hop* | *Funk* | *Groove Rock* |

| Watch | Listen | Tab Version | Print | Notation Key |

Figure 5.3
One of the first drum beats posted at OnlineDrummer.com (OnlineDrummer.com 2004). (*Screenshot by the author*)

(discussed later in this chapter). Like Next Level Guitar, OnlineDrummer hosts a thriving messageboard-style community forum; at the end of 2010 it had over 12,000 registered members, with about 810 daily visitors averaging 168 new posts per day (OnlineDrummer.com 2010a).

VIRTUAL MUSIC LESSONS

When I interviewed David Taub, Nate Brown, and other online music teachers about their work, I tried not to frame my questions in terms of the pros and cons of online versus face-to-face lessons. Instead I would ask something like "How would you say that video lessons are different from face-to-face lessons?" Nevertheless, teachers seemed primed to respond in a pros-and-cons format:

KM: Can you imagine from the student standpoint, what the key differences would be between face-to-face—

NATE BROWN: The student standpoint, there are some—I don't know if you want advantages or if you have another question on advantages—

Online music teachers come prepared to discuss pros and cons in part because they are constantly trying to anticipate their students' perspectives. Like radio show hosts, who strive to cultivate an imagined relationship with individual listeners who are unknown to them, online music teachers have to work at imagining the specific questions and concerns of the students on the receiving end. This is a major shift from the intimate social dynamics of face-to-face instrumental lessons (Kingsbury 1988, Nettl 1995).

Teachers' explanations of their own work are also conditioned by media coverage, as with the video-game reception discourses discussed earlier. For instance, when WNYC radio's "Soundcheck" program covered online music lessons, the producers chose to make this topic the subject of a "Smackdown" segment (a recurring element of the show). The segment began with the sound of a boxing match bell, and the guests were required to start by assessing online music lessons on a "pass/fail" basis:

JOHN SCHAEFER (HOST): As long as humans have made music, they've taken lessons. But increasingly the ritual of sitting beside a teacher and learning an instrument has given way to online alternatives. . . . So today, we'll grade online music lessons: Do they pass or fail? . . . [boxing bell chime] To you, Edward, first, our question of the day: pass or fail on online music lessons?

EDWARD SMALDONE (director of the Aaron Copland School of Music at Queens College): It's one of those questions that doesn't have a definitive answer, I think. There are—

JOHN SCHAEFER: I'm pushing you into a corner, Ed, pass or fail?

EDWARD SMALDONE: Well, it passes, but marginally. . . . You can learn things. But I think there's a wide gulf between learning how to play the opening lick of "Honky Tonk Woman" and being able to perform the Hammerklavier sonata. (WNYC 2010)

Smaldone's "Smackdown" debate opponent was Jeff McErlain, a guitarist who produces prerecorded video lessons and offers live video-chat lessons via Skype. As one might imagine, this opening immediately put him on the defensive. Because online teachers must constantly explain, promote, and defend their work, they are accustomed to discussing the difference between online and in-person lessons in a particular way—often along "Which is better?" lines.

Jeff McErlain, David Taub, Nate Brown, and other online teachers I interviewed all highlighted the same basic advantages of online lessons: convenience, repeatability, and affordability.[3] These are the same qualities that underpinned the explosive popularity of recorded music in the early days of the phonograph, which displaced some amateur music-making at home and music pedagogy at many schools. Advertising drove home the idea that the phonograph "fostered the appreciation of 'good music' and could do so for all Americans, regardless of ability, wealth, or location" (Katz 2004:55). Online music teachers employ similar rhetoric to explain how their lessons might reclaim domestic space for amateur music-making:

DAVID TAUB: The students can watch the lessons any time, and aren't locked into being somewhere between six and seven on a Thursday night. The lessons are loaded onto my servers. So it's like you can watch them at three in the morning or at three in the afternoon. You can rewind them, you can go back to them, you can make notes. I have a checkmark system, where after they watch a certain amount of a lesson, it gets checked off, so they can keep track of their progress. It's very convenient; they don't have to leave their house. You know, everybody's busy in today's world, and they don't have a lot of free time. And it's less expensive than private lessons. So there's some definite advantages to the online learning, for sure.

David mentioned that many of his students are parents; they don't have the time or the schedule stability to arrange for a weekly private lesson, but they can pick up the guitar and do a quick lesson after the kids are in bed or between loads of laundry. The "convenience" factor also applies to geographic constraints; as Nate Brown reported, "I get a lot of emails from people that say, 'Hey, I live out in a remote area and I don't have anybody that can teach me,' or 'I can't find a good teacher.' They thank me—'I'm glad you're doing this because this is all I have.'"

While this emphasis on *convenience* sheds light on the practical appeal of online lessons and the way that they fit into students' everyday lives, lesson *repeatability* draws attention to the distinctive pedagogical qualities of prerecorded video lessons. Of course, students sometimes make audio recordings or videos of their traditional private lessons, with the encouragement of their teachers. It's certainly educational to review one's own performance in a lesson and to hear how the teacher's guidance can elicit a different sound on the spot. But these recorded lessons aren't designed to be repeated or to compel active engagement; taking an hour to review a recorded lesson can feel like sacrificing a more productive hour of practicing. By contrast, online lessons are designed and edited to be effective interactive learning tools. Camera angles and close-ups channel the student's attention, and they require teachers to decide in advance exactly where that attention should be focused. Online lessons employ the "modeling-and-imitation" method typical of face-to-face instrumental lessons around the world, but the prerecorded video format allows students to control the pace and quantity of repetition (Campbell 2004:10–11). I can watch Nate Brown play a beat pattern, then pause and try it myself (twice or a hundred times), then back up and freeze-frame to try to figure out how he is coordinating his hands or gripping the sticks, then play the section again and see if I can keep up with his pace. A prerecorded teacher is infinitely patient. I can repeat two minutes of a lesson for an hour without feeling like I'm squandering the fee for a weekly lesson.

I discussed this characteristic of online lessons with Jeff Thompson, a guitar student who started learning with the Next Level Guitar curriculum but switched to taking private lessons with David Taub when he discovered that they lived in the same city. He explained,

> [David] films his hands up close; you can watch it over and over and over again until you get it right. The personal lessons, you can only get him for an hour, and then you gotta reschedule. My problem with the personal lessons is we go over so much material, and then I have to come home and I try and go through it as much as I can so I don't forget it right away. With the online stuff, you don't have to worry about that. You just fire the video back up.[4]

Although Jeff prefers the instant feedback and personal mentorship afforded by the private lessons, he continues to exploit the "repeatability" advantages of video when he is trying to learn new songs:

JEFF THOMPSON: The first thing I'll do is I'll try and find the actual band playing the song live and see what they do, like on YouTube or something, and I'll try and see just what the guitarist is doing. And then I'll just look for other people who are doing "how to play this song" or "how to play this solo" [instruction videos]. . . .

KM: What are you getting from watching it as opposed to just listening to the recording?

JEFF THOMPSON: What I'm really not good at still is just being able to pick up stuff solely by ear. So what I'll do is I'll watch the band play it live, and I'll just try and watch the guitarist and where he's at—like if he's playing a certain solo or certain section. If I just watch him, I can pick up what position he's in on the guitar.

From the teacher's perspective, video repeatability means relief from drudgery. While I repeat the shift between two chords a hundred times, study David's hand position, and take comfort from his repeated reminders that the F chord is really tough for everyone, David can be recording a new advanced lesson, meeting with his web-developer partner about new business strategies, driving down the California coast for a performance gig, or replying to questions on the web forum. Now that the beginner lessons have all been made and posted, he'll never have to spend an hour teaching someone the fingering for three basic chords again. As Nate Brown told me,

> When I was teaching private lessons I had about 44 students a week lined up, and I would teach the same lessons over and over and over. After a while, as a teacher you get really good with the fundamentals and what you're teaching— you really learn it—but it gets monotonous, and it can get to you. A video lesson, you teach it once, you can edit it, make it exactly as you want it, and it's out there and it's done, unless in the future you want to update it. So I think you can do a lot more material, because you teach it once and then you move on and you can have them all out there. . . . Also the teacher has a lot more time to develop a quality lesson and present it exactly the way that he wants.

Creating a video curriculum frees teachers from explaining the same basics over and over, while also freeing students from working with bored and apathetic teachers. Like video-game designers at work on the initial levels of a game, online teachers devote a lot of energy to making their earliest lessons as compelling and pedagogically effective as possible; excellent entry-level lessons will earn them loyal, long-term students.

Teaching technical fundamentals is one of the most difficult things to do in the online format, because students are still cultivating a physical relationship with their chosen instruments. This is why David spent 11 minutes explaining how to hold a guitar pick, zooming in with the camera and encouraging me to get to know its dimensions, angles, and texture. As an online commenter noted in response to a blog post I wrote about this lesson,

> The medium of the video promotes the translation of haptic knowledge into verbal and visual discourse (11 minutes on holding a guitar pick). While the teacher

can't mold your hand or shove your elbow up and down (ah, the memories of high school violin lessons with a Moscow Conservatory–trained chain-smoker!), there's no scarcity of time or of words. . . . A superabundance of time, visuals, and words replaces the experience of a grumpy tobacco-scented man yanking your elbow for 45 minutes while saying, "No, the WHOLE bow." (Cecire 2010)

But while students can take as long as they want to work through a lesson, teachers must still make careful choices about content, balancing numerous competing demands: attracting views, appealing to beginning students, building a sustained following, and working in accordance with their own teaching ideals.

FUNDAMENTALS VS. FLUFF: CRAFTING A CURRICULUM, BUILDING AN AUDIENCE

David Taub and Nate Brown spent years as students and teachers in face-to-face private lessons before they began their online teaching. They both stressed that private lessons are still the best way to learn an instrument, if students have the time, the money, and access to high-quality instruction. Accordingly, they imagine their constituencies to consist mostly of students who *lack* this time, money, and access—so they want to make their lessons time-efficient, free or cheaper than private lessons, and of the highest possible pedagogical quality. David believes that his online curriculum can be more effective than run-of-the-mill private guitar lessons:

> I still think private instruction is probably the best way to learn, *if* you have a good teacher with a good, structured plan. I think that's the fact. But I think online is a pretty close second. . . . And I wanted to teach more people. I wanted to bring my methodology to people all over the world. And I can't do that privately.

Nate Brown is more circumspect in his assessment of online lessons. (It should be noted that he is also younger, a less experienced teacher, and less professionally invested in his website than David Taub.) He told me, "It's a great tool, but I don't think it should overcome or take the place of private lessons. I think it's a good supplement or an opportunity for people who can't get private lessons. I think you could learn a lot from it. I don't think it's the best way."

Both teachers expressed distaste for "cookie-cutter lessons" and suggested that many private teachers fall into this mode when teaching beginning students. David told me that when he's getting started with a new private student, he always takes 90 minutes to discuss the student's specific guitar goals: "What do they like about music, and where do they see themselves on the guitar? . . . I tailor their lessons for their individual goals." This approach is obviously at odds with the constraints of a standardized online curriculum, which is "cookie-cutter" by definition. So David compensates by trying to make his videos feel as personal as possible:

> I try to make it very comfortable for the student, a comfortable atmosphere where it's just two people sitting in a room learning or playing. And a lot of my students online really relate to that. They say "David, I'm in—wherever—I'm in Canada, but I feel like you're right in the room with me teaching." . . . There was a time early on when I tried to slick the videos out a little more, I did some green-screen work and tried to put slick backgrounds in and make it really polished. I got a lot of emails about that. And everyone, they were pretty much, like, "Oh, man, put it back the way it was before."

As David's student Jeff Thompson told me, "Obviously there's no physical contact, but I can tell just the way he starts every lesson—he's always staring into the camera, and he's always directing towards the student, and I think he probably goes well out of his way to make sure, to try and make some kind of connection through the camera that way."[5]

No matter how intimate and charismatic the teacher's screen presence, pre-recorded video lessons do not allow for instant feedback on a student's performance. This situation has its advantages; for instance, it totally eliminates the performance anxiety that many students associate with private lessons. My teacher is not assessing my talent or my progress. As Henry Kingsbury notes, in traditional music pedagogy contexts "the 'environment' of a person's musical development must be understood in terms of personal relationships characterized by some degree of intimacy and vulnerability. . . . An authoritative negative judgment tends to become a proscription against subsequent musical-social behavior" (1988:72–73). But if I am learning from video lessons, no one is going to brand me as fundamentally "unmusical" or untalented; indeed, teachers like David Taub and Nate Brown are invested in persuading viewers that absolutely anyone can learn to play. I have only myself to please, and I don't have to prove how much I've practiced in the week since my last lesson. Why be nervous about my performance when I'm alone in my living room?

However, the lack of instant feedback also makes it possible for students to develop bad habits that might have to be "reverse-engineered" later, as David put it. Jeff Thompson explained what happened when he switched from the online curriculum to private lessons:

JEFF THOMPSON: I didn't have good technique on how to keep my hands tight against the guitar so that I was being as efficient as possible with changing chords. That really comes into play when you start playing more difficult, faster stuff. . . . [David] picked up on that on the very first lesson. He said,

"I want you to work on when you change, make sure that your hands are like this, don't pull your hand away from the guitar, otherwise you're not gonna be able to get back to it in time."

KM: In the videos, David also really emphasizes that point, staying close to the fretboard. So what was he able to do in person that really made it work?

JEFF THOMPSON: He basically just forced me, so we'd be playing something or he'd have me play some kind of progression that would show that off if I wasn't doing that correctly, and he would just be, like, "Nope, start over." He'd push my hand against the board and be, like, "You need to not move it more than this." In person, it was the instant feedback right away, whereas the video, I'd be watching something, like "Yeah, okay, I kind of get that," and so I wouldn't keep focused on it as much.

The online teaching context also eliminates an important motivator for teachers: They can't track the efficacy of their teaching. Nate Brown and David Taub both identified this as one of the key challenges of online teaching. As David told me, his greatest satisfaction lies in seeing students "get a little bit better at a time, and I see their progress and how happy it makes them." To keep their work meaningful, online teachers have to work to maintain a sense of a personal connection with individual students—both by imagining a virtual student's perspective and by encouraging students to make contact via comment threads, web messageboards, or email.

Nate Brown and David Taub clearly believe that online lessons represent a compromise, not a high-tech triumph that will inevitably replace private lessons. This knowledge motivates them to think carefully about how to make the best possible use of what they consider to be a second-best pedagogical format, as well as how to compensate for its shortcomings. For both teachers, this means a strong emphasis on structure: a graded curriculum that helps students develop fundamental skills, delivered with carefully calibrated "instructional timing and flow" (Campbell 2004:4) that provides new information exactly when it is needed (Gee 2004:65). The online format is well suited to presenting a standardized curriculum; lessons can be presented in a particular order, and a database can help students track their progress. A structured curriculum also lets the teacher assume that students have already acquired particular techniques and knowledge by the time they reach a particular lesson. As Nate Brown explained, it's very difficult to plan a "one-size-fits-all" lesson: "I'm teaching a lesson on how to play this song. I don't know if someone listening already knows what a sixteenth-note is. . . . I have to decide, am I going to go out on a full explanation, or how deeply am I going to go into this?" David Taub repeatedly emphasized that structure lies at the heart of the "Next Level" concept. Work through the lessons in order, moving from one level to the next, and success is guaranteed:

A lot of students get turned off early on because of either bad teachers, or they're not following a structured plan, or they have unreal expectations, and then they put the instrument down. It's a shame. Because if they had a good teacher who was able to really set up a structured curriculum and teach them how to practice, they can play. I can teach anyone how to play—anybody.

Both Nate and David decided to begin from the beginning—not just in arranging the order of their lessons, but in actually creating them. For David, this meant imagining a student "who comes in to me and says, 'I've never played guitar in my life. I don't know anything about it. I don't even know about a guitar—what one to buy, should I play acoustic or play electric. I have no clue what to do, but I know I want to do it. How do I get going?'" (David's curriculum also illustrates his steadily advancing production skills and equipment; compare ⏵ 5.1 to a more advanced lesson, ⏵ 5.5.) Nate began by creating a "Learning to Read" series—"quarter notes, whole notes, basic music theory"—because he believes that "learning to read music opens up a lot of opportunities for you. You can play in a studio, you can read charts, jazz bands—you're going to get more work, not to mention you'll be able to learn from a lot more people that have written books." (See online-drummer 2007, ⏵ 5.6.)

However, this emphasis on fundamentals potentially conflicts with the need to garner a high number of page views, especially on YouTube. Potential students are often looking for flashy techniques or lessons on how to play particular songs, fills, or solos—recall that Nate's first video was created in response to a request for a "Back in Black" lesson. When Nate started posting his "Learning to Read" series on YouTube, he initially disabled the star-rating function:

> When I started I had to make a decision—[should I] teach lessons that are exciting for people and make them happy, a lot of the fluff? Which would have made me, I think, more popular to start. A lot of guys do that. They teach, "Oh, here's a cool fill," or "Here's this part of this song." I decided that to start I wanted to teach the fundamentals. . . . I thought at the time it's probably not going to receive that high of ratings from the average user looking for drum lessons, but I think it's going to be very helpful. And if the ratings were lower because of the people that are looking for different lessons, it might ruin the credibility for people who really want to learn it. They might see, oh, well, it only had three stars—but not because of the lower education quality. I didn't want them to overlook it because of that.

The challenges of impression management on YouTube suggest another reason why online music teachers might prefer to develop their own websites when they have the time, money, and skills to do so. It can be nervewracking to post a video and open it to public commentary, particularly given that vitriolic insults appear in virtually every YouTube comment thread. But Nate also noted that comments can bring up the ratings on a video, because "it makes it more interesting and people revisit it" as the comment thread develops. He eventually decided to allow ratings and comments on all new videos.

As they worked to promote Next Level Guitar, David Taub and Tim Gilberg also put a lot of thought into how to attract more views on YouTube. At one point they experimented with the venerable "hot girl" marketing principle. For instance, a video posted to YouTube on July 23, 2007, is advertised with a thumbnail image of an apparently naked woman embracing a guitar. The lesson itself is in the usual Next Level Guitar style, but halfway through the video the still image from the thumbnail appears for several seconds.[6] This kind of "Easter egg" is common in video games and online multimedia. On YouTube, Easter eggs are often referenced in a video's info section or comment thread, which encourages viewers to watch the whole video so that they can find the bonus material.

On the day this video was posted, the members of the Next Level Guitar online forum began to engage in a vigorous debate about the "guitar girls."

Thread subject: "Ditch the girls . . . "
Tim and David why are all the guitar girls popping up again? . . . Many of us are against the guitar girls but why are you bringing them back all of a sudden? Please remove them they get annoying when they keep popping up in the middle of the lesson.
–Valhalla, Member (profile icon: image of guitar body)

Come on man, who doesn't like the occasional hot chick?
–EVH5150, Senior Member, OKC [Oklahoma City], Oklahoma (profile icon: vintage-style image of Ronald Reagan with text "Old School Conservative")

I remember when it was asked, but I don't remember that many people complaining, I for one thought for marketing to all different people including different countries it would probably be better to drop them, but I read most of those posts and very few if any said that they were offended or anything like that.
–ez-one, Moderator, 7/23/07 (no profile icon)

Yeah, I don't want to be a prude, but I think you should get rid of the girls too. From a marketing point of view, are you trying to attract members through your talent (and you do have a LOT of talent) as a guitar teacher, or attract people by flaunting half naked girls that have nothing to do with your course/lessons? As a female, the girls don't really make me feel at ease and they almost make me feel left out.

To all the boys wanting to keep the girls, would you feel the same if the picture was of a man wearing a thong and holding the guitar seductively? I think you'd feel slightly ill at ease, too.

–vaxwell, Senior Member, Melbourne, Australia (profile icon: headshot photo, woman holding neck of guitar next to her face)

(Next Level Guitar 2007a)

Tim Gilberg posted the following response within five hours of the initial post:

Thanks for the Feedback, both Good and Bad

Let me share my thoughts with you and also We do take all this feedback into consideration

In order to gain exposure to the site I will experiment with various ads and techniques

In marketing there is a saying TEST TEST TEST
I see both sides let me share mine
When I go to barnes and noble and view the guitar magazines on the shelf
In many cases there is a very HOT Bikini clad model etc on the cover
So I am not introducing anything groundbreaking here

Having a sexy model pose with a guitar is far different than banning profanity on this forum

YouTube is our marketing tool
Chicks = clicks is a phrase i coined and it does work
Now many of you have been with us since the beginning and you know and respect David's methods

Let me ask you this if the picture draws interest to our site and then they see and hear David and go WOW this guy can teach

MISSION accomplished
We recieved 59,000 clicks on the video I added the first girl to in less than a week many other videos receive well under that
I of course value our members but having a sexy girl with a guitar I hope you do not view as profanity

This is an open forum and we are open to all ideas
ROCK ON
Tim

The debate continued, with most forum members arguing that the images made the videos look less professional, distracted students from the lessons, and might be off-putting to potential new students. The following afternoon, Tim posted this message:

> Hmmm looks like i was wrong
> It looks to me that my test was not approved by the good people
> that is the beauty of these forums,
> So i will not be adding any more girls to the middle of the videos

This discussion, which took place in the first year that Next Level Guitar was open for business, demonstrated that the site's student population was more diverse than many people had realized—in terms of both the participation of female students and the variety of students' cultural norms about gender and sexuality. The YouTube lessons and website were apparently reaching a different audience than those guitar magazines with bikini-clad models on the cover. Three years later, David told me that he had deliberately recruited some female guitarists as instructors for his lesson DVD series in an effort to reach more female students:

> I think it's important to show that women can be just as good as guys. Guitar is, like, 95 percent male. It's so male-driven. I really want women to know that if you want to play a guitar, you can do it, too. So I have a few women teachers that I use. And it's interesting that since I've been using them, I mean, more and more have been signing up on the site, so that's great. . . . Because [guitar] is just such a male thing—such a masculine thing—and it doesn't have to be.

I also discussed the gender-imbalance issue with Lisa Tye, an Australian stay-at-home mom who is a volunteer moderator for the OnlineDrummer community forum. She acknowledged that "a lot of music forums tend to be male-dominated, especially drums, as they are perceived as more of a male instrument." However, she told me that at OnlineDrummer, "We aren't treated any differently to any other member because of our gender. In our small group of moderators, there is myself and one other female, which I really think helps members see it's not a judgmental community."[7]

MAKING ENDS MEET AND/OR GETTING PAID

Because the upsurge in online music lessons coincided with a major downturn in the U.S. economy, many media stories about online pedagogy have focused on its free-or-cheap nature. For instance, only a few weeks after Next Level

Guitar was launched, David Taub was featured in a story on National Public Radio's *All Things Considered*:

FRANK LANGFITT (REPORTER): Let's say you want to learn to play guitar—but you don't have the time or money for lessons. Why not try YouTube? A number of people teach guitar on the video-sharing website, offering lessons for free. In the past few months, two teachers have posted around 200 videos that demonstrate everything from basic strumming techniques to the opening riff of "Sweet Home Alabama." So far, people around the world have watched the videos a total of more than 3.5 million times.

One of the teachers is David Taub, who lives in San Diego and often appears wearing a flannel shirt and a backwards baseball cap. A one-time bar-band rocker from New Jersey, he opens each video with the same line: "What's up, good people!" . . . Taub's lessons are mostly unedited and include moments like his golden retriever eating his guitar pick. (Langfitt 2007a)

As the story continues, the fact that Taub charges for lessons at Next Level Guitar seems to undermine the charming folksiness of this introduction.

On the free videos, Taub teaches the basic chords to popular songs, but he holds off explaining some of the riffs so he can drive people to his site. After playing a riff from Sheryl Crow's "If It Makes You Happy," he stops playing and says, "But if you want to learn that, you're going to have to go to our full site for the lead lines, okay?"

The reporter juxtaposes David Taub with Justin Sandercoe, another very popular online guitar teacher who does not charge for lessons:

Justin Sandercoe also has a teaching website—justinguitar.com. He has a few ads and takes donations through Paypal to cover the site's hosting fees. But Sandercoe doesn't charge visitors; he says he sees the site as more of a public service. "I like the idea of being able to deliver quality guitar lessons to people who can't afford lessons, or who are in places where there's not that kind of access to somebody who can teach them the right stuff," he says.

When Sandercoe was growing up in Tasmania, it wasn't easy for him to find great teachers. He hopes his videos will help kids in places like Sri Lanka or India who may not be able to learn otherwise.

Like much media coverage of this phenomenon, the NPR story implies that online music lessons are worthwhile mainly as a charitable service to impoverished or not-yet-committed students, or perhaps as a fun supplement for people who are already taking private lessons. (The story closes by assuring listeners that in-person lessons are still the best way to learn.) But why shouldn't online teachers make a living from their work? Like Nate Brown and Justin Sandercoe, David Taub does post quite a bit of free material to YouTube. As he told me, "It's a nice way for students who can't afford lessons around the world, and also it's a good way that I can introduce students to the program." But David works full-time producing and maintaining the Next Level Guitar site. Server space, video equipment, and video-editing software all cost money. Why should his lessons be free?

One pragmatic answer was foreshadowed in the NPR story: because teaching people popular songs may constitute copyright infringement, and those who charge for lessons are most likely to be targeted.

> If learning pop songs for free online sounds too good to be true, it may be. John Palfrey, executive director of the Berkman Center for Internet and Society at Harvard Law School, says most of the songs Sandercoe and Taub teach are under copyright. He thinks it's only a matter of time before a licensing company orders YouTube to take them down. "There's a very strong argument that the reuse of well-known chords in the sequence the instructor played them would be a violation of the copyright," Palfrey says. (Langfitt 2007a)

Indeed, in a follow-up story three months later, NPR reported that YouTube had removed David Taub's videos in response to a copyright infringement claim. The complaint focused on a lesson that taught students to play the Rolling Stones song "Brown Sugar," but YouTube shut down the whole channel. Sandercoe's channel remained available:

FRANK LANGFITT (REPORTER): Interestingly, you can still watch videos from another big YouTube guitar teacher, Justin Sandercoe. What's the difference? Taub taught on YouTube for free, but he used those videos to market a paid teaching website called NextLevelGuitar.com. Sandercoe, on the other hand, didn't charge for any of his Internet lessons.
JUSTIN SANDERCOE: All of this stuff from my site is free, so therefore maybe people will see it more as a public service.
FRANK LANGFITT: Taub plans to post new lessons on YouTube, but he says they will feature original material. (Langfitt 2007b)

The NPR story also references a response video that Tim Gilberg posted to a new YouTube channel with the account name "rockongoodpeople" (the channel Next Level Guitar now uses). Gilberg plays and sings a brief protest song: "Explaining a few chords, talking about a song /Please let me know how that could be wrong . . . /We will start again, with the help of our friends. . . ." In the video's "info" section, Gilberg writes,

David and I are hard at work getting ready to bring you some Free ORIGINAL
video guitar lessons
You Can't Stop the ROCK
Thank you ALL for your overwhelming support
(rockongoodpeople 2007, see ▶ 5.7)

There are thousands of song-lesson videos available on YouTube today,
spanning "a spectrum of consumer co-creation that ranges from distinctly
communitarian to explicitly market focused" (Potts, Hartley et al. 2008:460).
At the end of 2010, a search for "brown sugar lesson" yielded dozens of guitar
lessons, some uploaded over two years earlier. Clearly, copyright enforcement
has been spotty. David Taub's channel was a victim of its own success; in some
sense, it is a badge of honor to attract the attention of copyright holders
(McGranahan 2010).

KM: Did you ever actually hit the point of getting a cease-and-desist order?

DAVID TAUB: No, no. . . . We got so big—huge—very, very quickly. And the
bigger people are always under the spotlight more. And it's a very gray area,
especially if you're putting stuff up for free, and are you benefitting from it?
Not just monetarily—are you benefitting by getting traffic to a different
site, or are you benefitting by getting students? Are you somehow benefitting from a piece of work? YouTube, if they get a certain number of complaints about something, they just shut your channel down. They don't get
involved in anything. So one of the publishers, ABKCO, they didn't like the
fact that we were teaching a Rolling Stones song for free. Just pretty ridiculous because it makes so many more people aware and some people go and
download the song. But they didn't like that, so they complained. And eventually YouTube closed our channel. It was so popular everybody just put
their arms up and went, "Oh my God." And that's when the media attention
got huge. . . . So after that we decided, okay, so we're not going to go into
court—we'll just buy the publishing rights. There was never any court.
There was never any cease-and-desist. It was all very informal.

David Taub found the media coverage of his copyright travails frustrating, because reporters often implied that he had been charging for access to song lessons on the Next Level Guitar website. In our conversations,
he emphasized that only his technical curriculum was behind a paywall; he
never charged for song lessons. However, both David Taub and Nate Brown
attribute much of their early popularity to their YouTube song lessons.
Many students want to learn to play guitar or drums because they want to
play particular songs, and both teachers continue to be inundated with song
lesson requests. Both have responded to this demand by creating instruction videos for particular songs—although in their lessons they try to convey

the message that songs are vehicles for learning new skills, not just ends in themselves.

Next Level Guitar now secures the rights required to publish song-lesson DVDs, which are sold through the website. These publishing rights do not include web broadcast rights, and David Taub no longer produces any song-lesson videos for the web (either for YouTube or for paying subscribers). Over two years after the copyright snafu, he seemed to have made peace with this episode and was happy with the sales from his song-lesson DVDs. He offers his subscribers deep discounts on the DVDs, and his students often share links to under-the-radar free song-lesson videos on the Next Level Guitar forums.

At OnlineDrummer.com and on his YouTube channel, Nate Brown continues to post both song lessons and sheet music for drum parts, painstakingly transcribed from recordings. He commissions some of the transcription work, since it is too time-consuming for him to do by himself:

> Originally I paid per 1,000 views of [the transcriber's] sheet music. So it would be—originally it was $2.50 per 1,000 views, which is a lot to pay because the advertising you bring in on that page is less. . . . [Later on] the transcriber got a Google AdSense account and I put his Google AdSense code on his sheet music pages so that Google pays him directly based on whatever money is earned. Basically I'm renting space on the site to this person, and he's being paid.

Nate Brown also sells lengthy tutorials for individual songs as downloadable videos through his DrumMusic.tv site, which was originally launched to help OnlineDrummer's core staff and volunteers fund a trip to the Percussive Arts Society International Convention. The tutorials are 25–35 minutes long; they include multiangle video examples and a PDF with full sheet music for the drum part. These lessons proved popular, and Nate continues to produce them to support OnlineDrummer's free content and to bring in some supplemental income for himself. He estimates that he spends 20–50 hours a week working on OnlineDrummer.com and DrumMusic.tv, in the evenings and weekends during the school year and full-time during the summer. He has thought about quitting his day job, but he worries about giving up a secure paycheck:

> With the way the economy is, you don't know. One day I can bring in a lot of money and the next day maybe not. Tomorrow somebody might invest all this money on a new site and there goes mine, or I could get hacked, which happened a few weeks ago, and the site's down, and then I have a family to support. So it's a big decision.

However, Nate Brown does have one significant strategic advantage over Next Level Guitar: Unlike lyrics, melodies, and chord progressions, drum parts generally aren't protected by copyright. This was an issue Nate had investigated in detail:

NATE BROWN: I have had people ask me about that before. We have a lawyer who is a moderator on the site and she [has done] some research into that, and I've talked to Phil Hood with *Drum!* magazine; they do the same thing. They publish parts of drum music and songs, DVDs, and with drum music you do not have to have the rights.

KM: So you've never gotten any kind of cease-and-desist? Because some of the videos do include, it sounds like, some of the original recording, right, because you're playing along with the recording.

NATE BROWN: That's where it gets sticky, and if somebody were to say, "You know what, that's our recording and you need to pull it," then I would. . . . I'm not actually trying to sell the song. The purpose of the video is to show the song played all the way through to give the drummers an opportunity to learn it. . . . If it went to the point where they're saying, "Hey, you're costing us money because people are just buying your lesson to get the song," and we went before a judge, they would have to determine what the intent was and if there's any damage being done, which I don't want to do any damage.

Nate Brown is alluding to a possible "fair use" defense, based on both the educational character of his videos and their presumed nondamaging effect on the market for the copyrighted works.[8] Meanwhile, OnlineDrummer and DrumMusic.tv are relying on "a kind of default slipping through the cracks," as Nate put it, based on U.S. copyright law's privileging of words and melody over rhythm.

AUTHORITY, AGENCY, AND COMMUNITY

> David talked to the cameraman occasionally, requesting zoom in/out etc. It became clear that he could see himself on a monitor. Maybe this helps him maintain eye contact, which is a really compelling part of the videos. On the other hand, I still felt free to multitask. In Lesson 12 he spent a few minutes telling people not to be afraid of music theory, which wasn't really key information for me, so I messed around plucking the strings and reviewing their names. I also had my email tab visible during the lessons and at one point during a nonessential moment I switched over to see what message had just clicked in. I felt freer to experiment physically and generally fidget with the guitar and pick than I would if a teacher were right in the room with me. I think then I'd feel like I should just sit still and pay close attention until asked to do something in particular. (fieldnotes, December 9, 2009)

Online lessons destabilize the teacher's authority. While David worked to direct my view, actively managing camera angles and close-ups, he couldn't tell whether I was following directions or even paying attention. Sometimes my multitasking was productive, but the temptations of other online pursuits were difficult to resist. I see this same scenario play out in the classroom all the time: Students with laptops do web searches on course-related topics (or not), or they send text messages under the seminar table. Like many of my colleagues, I find these practices frustrating and disrespectful. After all, I'm *right there*, and students have a limited number of minutes to benefit from my class prep and their classmates' contributions to discussion. However, I felt only a ghost of guilt when I checked my email during my guitar lesson. David's labor was already done and paid for; he could be doing anything right now. If I missed something important, I could just play the lesson again.

It's different when I'm playing a song in *Guitar Hero* or completing a time-sensitive mission in *Grand Theft Auto*. The games constantly evaluate my performance and provide feedback, which helps make them completely absorbing. While I can always hit the Pause button, doing so feels like a major interruption rather than casual multitasking. Yet I'm reluctant to call the games "more interactive" than my online guitar lessons; instead, they channel my attention differently and offer me incentives to stay focused. These contrasting experiences offer a reminder that in virtual performance contexts, interactivity is a state of mind (Lee, Park et al. 2006; Gonzales, Finley et al. 2009). David Taub believes that "a big part of teaching is listening to the student, and what struggles they're having with the instrument"; when he makes a video, he has to imagine those struggles. Students in turn do the imaginative work of interacting with the teacher—but they have to commit consciously to doing that work. Jeff Thompson, the student who worked through the Next Level Guitar beginner series before switching to private lessons with David, told me that he pushed through the videos in much the same way that he would approach the levels in a video game or a series of songs in *Rock Band*: "I had this sense of just wanting to get through them . . . to get to the next one, to get to the next one, to see what the next one is, since that's how you finish the game."[9] When he switched to private lessons, he had to turn over more control to David, who could say, "Nope, start over!" or actually push his hand closer to the fretboard.

Since online teachers can neither confirm that students are following their instructions nor reprimand them for straying, it behooves them to adopt a persuasive rather than a dictatorial teaching style—for example, to explain *why* it's important to keep your fingers close to the fretboard. David Taub and Nate Brown also try to model an open-minded and flexible approach to instrumental technique; when possible, they offer students multiple technical options and encourage them to explore their own preferences. This pedagogical style encourages engagement, since students have to consider their options and make a decision. For instance, in his basic samba lesson Nate Brown says, "A lot of people like to break it down by the hands first [*demonstrates hands only*] and then add the feet in. And you can do it that way. For me, I like to learn it all together,and just look and say, 'Alright, on count 1 I'm doing this' [*demonstrates*], and take it little piece by piece. But whatever your learning preference is—you know, give both a try, figure out what *your* learning preference is" (OnlineDrummer.com 2010d).[10] David Taub offers similar technical options for the guitar, and he also exhorts students to consider the same questions he'd be asking them in a private lesson: "What are your ultimate goals? Where do you want to see yourself with the guitar?"

In another move toward dispersed and distributed teaching authority, both David Taub and Nate Brown have embraced the amateur-to-amateur pedagogy potential of the online format. When I'm taking a lesson, at any moment I can drop the "alone in my living room" approach and seek advice and encouragement from other students on the web forums, or I can do a YouTube search to see how other guitarists hold the instrument or the pick. If I want an audience for a song performance or tips on my technique, I can post a video to the forum sections designated for this purpose. The Next Level Guitar forum promises a friendly audience: "If you just started to learn to strum 2 weeks ago or applied something you learned from David's Intermediate advanced or a video song lesson we would love to see it. Any haters will be banned immediately in this section." There were almost 2,000 posts in that section by the end of 2010, offering a wealth of crowd-sourced feedback.

While online teachers often crave feedback from students, they also face the challenge of responding to an avalanche of questions, comments, videos, and personal messages if they gain a large following. For teachers with hundreds or thousands of students, online messageboards play a huge role in reducing this burden and in creating a sense of "community"—the term that designates the forum section at both Next Level Guitar and OnlineDrummer. Nate Brown told me that the forum is "the most important part" of OnlineDrummer—"It makes people feel ownership and a part of something, it allows people to actually have that other direction," in contrast to the one-way communication of a prerecorded lesson. Students post questions that are already addressed in video lessons or printable tip sheets, and they often receive the same answers already supplied by those sources. Despite the redundancy, forum members clearly find this process satisfying; getting individual responses rather than bullet-point tips assures students that someone cares about their progress with the instrument. Many of the posts that garner high numbers of views and responses seem to invite community support as well as technical advice, for example, "How do you 'just jam'??"(OnlineDrummer.com 2010b); "I HAVE AN AUDITION AT DRUM-TECH IN LONDON. I'm nervous" (OnlineDrummer.com 2010c); "Fellow Beginners, Show youselves!" (Next Level Guitar 2007b); "Just learned my first real solo" (Next Level Guitar 2008).

From a business perspective, what's happening on these forums might be described as outsourcing customer service to improve efficiency and teaching-load capacity. From a teaching-and-learning perspective, the forums constitute virtual communities.

Values-oriented discussions like the Next Level Guitar "Ditch the girls" thread show that online forum participants take seriously their own status and responsibility as members of a community. Such forums potentially enable virtual versions of the "studio" or "lineage" model for traditional private music lessons, in which students are affiliated with particular teachers and feel a special connection to their teacher's other students (Kingsbury 1988:37, Nettl 1995:69). But in the case of Next Level Guitar, students are also conscious of their status as customers—and David and Tim seem comfortable addressing them that way, rather than demanding the respectful deference due to an authoritative teacher. For a web entrepreneur, getting detailed feedback about marketing strategies like the "chicks = clicks" approach is priceless. For the forum members, Tim's response to their concerns offered satisfying confirmation of their status both as a community—one in the process of developing its own social norms—and as a valued customer base.

A VISCERAL INTERVENTION

Even before the Internet age, "self-taught" instrumentalists never learned in a vacuum. Aspiring musicians have long learned from recordings and live performances by other musicians, as well as by comparing notes with friends or the guys (and they are notoriously mostly guys) down at the local music store. I borrowed my acoustic guitar from a friend, an accomplished musician. After about a week he stopped by and wanted to see my progress.

> It was amazing how quickly I learned a couple of things just because J. was able to tell me that what I was doing looked really weird to him. We established that I could hold the guitar much better if I crossed my legs and propped it on top of my right thigh, and that I didn't need to hold it so tight/parallel to my body. I showed him my bad up-strum and he didn't say what specifically looked weird about it, but then when he demonstrated I saw right away that his hand moved in a diagonal across the sound-hole (which I could have inferred from the shape of the pick guard), rather than straight up and down. I had been trying to move the pick perpendicular to the strings, against the natural arc described by my arm moving from the elbow. My up-strum hugely improved right away.
>
> I wonder if David ever gives this kind of feedback based on watching videos of students? Obviously if I were really mirroring David I would have figured out the strum arc on my own, but since I couldn't see myself I couldn't tell that what I was doing looked weird.

I feel like I'm gaining a very different concept of the geometry and ergonomics of playing guitar. Piano, clarinet, and even leading Sacred Harp tunes seem to operate more in two dimensions or along perpendicular axes (relatively speaking). With guitar, it seems like individual finger joints have their own rotational axes. I was also very self-conscious about playing in front of someone. It didn't quite feel like going in for a piano lesson, but it reminded me of that feeling—have I prepared enough for the teacher? (fieldnotes, December 17, 2009)

My friend said he had wondered if I would be strict about just using the video lessons—a restriction that might have lent my virtual fieldwork a patina of old-school empirical authority. Would it ruin my experiment to learn from other sources? Or, switching to a new-school, ethics-oriented ethnographic paradigm: Would it be disloyal to my teacher, or at odds with the values of the community I was joining? David's and Nate's own comments and thousands of student forum posts convinced me otherwise. After all, the whole point of these online lessons is to bridge the gap from virtual to visceral, from theory to practice, and from imagined musicianship to actually playing. On the forums, students constantly encouraged each other to take their playing public.

CHAPTER 6

Amateur-to-Amateur

Shouldn't we all be playing our games and taking our music lessons on the holodeck by now? I'm writing in 2011, a year that signified techno-utopia (or dystopia) when I was a kid reading classic science fiction in the 1980s. Totally immersive virtual reality has been right around the corner for decades now. In 1996 a prominent drama scholar confidently predicted, "By the year 2010, the perfection of virtual reality . . . will have added unheard-of dimensions to the field of performance" (States 2003 [1996]:125). But instead of downloading basic guitar skills straight to our brains and moving right on to playing awesome solos, lots of us are sitting in front of a screen and laboriously strumming along with a video, then hitting the Replay button until our fingers hurt too much to continue. We still have bodies and we still have to practice, whether we're playing *Guitar Hero* or a guitar with strings. Moreover, we still have to learn from *other people's* bodies, finding a way to comprehend their kinesthetic knowledge and make it our own. Our current online media formats seem ill suited to this purpose: two-dimensional, with a radically impoverished sensorium (just sight and sound), and often lacking real-time interaction. Nevertheless, body-to-body transmission is taking place online, as millions of people turn to YouTube, blogs, and other web-based social media in the pursuit of new corporeal skills, experiences, and knowledge. As the previous chapter demonstrated, many experienced teachers are ready to assist them. But there are far more fellow amateurs: documenting their own learning process, eager to compare notes with others, or even offering tutorials, despite their limited expertise and lack of formal credentials.

Amateur-to-amateur online learning is part of a vast array of contemporary peer-to-peer online interactions. "Peer-to-peer" (P2P) was originally a technical term that described distributed computing architectures in which tasks and resources are shared across a network of computers (Khan and Wierzbicki 2008). The term entered common parlance in connection with the mp3-trading service Napster (established in 1999 and shut down in 2001 after legal battles with the recording industry). "P2P" became synonymous with file sharing, especially illegal traffic in copyrighted music or videos. In the years since, the P2P concept has been adopted to describe web-based communication practices. Rather than going through a higher authority (i.e., a multinational corporation or an accredited expert), online "peers" cut out the layers of middlemen and share their resources directly. I propose "amateur-to-amateur" (A2A) as a useful term for discussing certain kinds of P2P teaching and learning. It retains P2P's emphasis on destabilized, dispersed authority while foregrounding practice and process: Amateurs *do* things. A2A practices ultimately aim to share kinesthetic knowledge, not digital files. "Amateur" also draws our attention to motivation and affect: Amateurs do things *because they want to*, for pleasure and at their leisure, not for material compensation or under duress. Finally, amateurs always imply the existence of professionals or accredited experts, just as the "peers" of P2P imply the existence of higher authorities (Stebbins 1992:38–58).[1]

In this chapter I frame three windows into the contemporary A2A phenomenon, analyzing online conga drum lessons, popular-song piano tutorials, and ashtanga yoga videos and blogs. Each of these practices has its own complex history and established transmission methods, which I will only treat allusively here; my aim is to present concise case studies that lend themselves to comparative analysis. I have chosen this approach in part because of my experience with student projects of similar scope. Nearly every course I teach involves individual ethnographic research projects, documented on student research blogs. The resulting collection of multimedia mini-ethnographies always generates unexpected connections and comparative insights in class discussion—itself a demonstration of what amateur-to-amateur pedagogy can accomplish. I therefore offer this chapter as a gesture of appreciation to my students, in the hope that it will usefully illustrate both the potential and the inherent limitations of comparative mini-ethnographies.

FINDING A CONGA TEACHER ON YOUTUBE

Imagine that you've recently become a big fan of salsa. You decide you want to learn more about how the music works—maybe even take a few conga lessons yourself. On a whim, you do a YouTube search for "conga lessons." Today the top result is a lesson called "Poncho Sanchez—Fundimentals of Latin Music—Conga," posted by Gordanius in 2007. It has attracted 563,176 views, which seems to lend it a bit of authority. You hit Play, and Poncho Sanchez introduces himself and starts to explain the names of the different-sized conga drums in front of him. (See Figure 6.1, ▶ 6.1.) While the video runs, you

Figure 6.1
Still from a Poncho Sanchez conga lesson on YouTube (Gordanius 2007). (*Screenshot by the author*)

idly scroll down to the posted comments. In the "Highest Rated Comments" section, someone is defending Sanchez against a criticism that's no longer onscreen:

> mexicanbarracuda: azcaine69 nada mas la pura pinche envidia de que su sangre no tiene el calor latino pinche gringo pendejo por mucho que estudies nunca vas a lograr igualar el toque latino
> con eso solo nosotros nacemos
> no hay nada mas chingon que ser latino!!!!
> (comment posted on Gordanius 2007)

If you can read Spanish, you understand that mexicanbarracuda is accusing azcaine69 of attacking Sanchez out of envy. mexicanbarracuda's profanity-laden defense of Sanchez relies on the claim that no gringo can match the Latin touch on the drums ("only we are born with it, there's nothing more badass than being Latino!!!!"). Further down the page, another commenter criticizes Sanchez—first for pedagogical inefficiency (he could explain the same material in fewer words) but also for playing that lacks "sabor" (flavor); he's great technically, but he doesn't *speak* with the congas.

> Joe0567: Uso mucho tiempo en hablar, pudo decir lo mismo en menos palabras [. . . .] Poncho es muy técnico, súper técnico y definido en sus ejecuciones, pero no tiene mucho sabor, es decir NO HABLA con las congas.

Another person recommends a different YouTube channel and notes that these days students don't have to rely on a single teacher's example:

> TiqueO6: check out lot's of pro conga players on martincongahead's channel (he's the head of LP percussion) and you'll be able to compare technique. Now we can see many examples rather than basing our concept on just one drummer

This seems like a good point, but you don't feel qualified to compare different players' techniques—you're looking for beginner lessons. You back up to your search results page and click on the second video in the list, which is titled "Conga Lesson 1: Basic Tones—Nate Torres." 85,052 views, not too shabby. Nate Torres introduces himself and promises that this video is the first in a series (prpapito3000 2007a; see Figure 6.2, ▶ 6.2).

Both the teacher and the video production seem far less professional than in the Poncho Sanchez lesson. The Torres video looks like it was made for YouTube, whereas when you Google the title of the Sanchez video you see that it's probably an unauthorized excerpt from an instructional DVD of the same name released in 2005. Sanchez is on stage or in a studio, with well-miked drums displayed from multiple camera angles. Nate Torres seems to be in his living room—or maybe more likely his parents' living room, given his age and the visible décor. The lighting isn't great and the sound reproduction is fuzzy. Torres fidgets with his hat and shifts in his seat; he has none of Sanchez's

Figure 6.2
Still from a Nate Torres conga lesson on YouTube (prpapito3000 2007a). (*Screenshot by the author*)

gravitas. On the other hand, by 35 seconds into the video Torres is telling you exactly how to position your hand in order to get an open tone out of your drum: "What you're going to do is have your fingers lined up just like this, close together, with your thumb out, so your thumb doesn't get in the way." He uses his own left hand to shape the position of his right hand, inviting you to do the same—a physical correction from an imagined teacher. He offers a visual cue for exactly where to strike the drum: "making sure that your knuckle line lines up with the rim of the drum." Then he slowly demonstrates three strokes and looks right at the camera, seeming to ask, "Did you get it?"

When you scroll down to the comments, you see that Torres must be "prpapito3000," the person who uploaded the video; he's responding to comments and questions.

> longislandsalsero: HEY BRO
> My Wife Got me some Matadors for xmas and i found your lessons. . . . they really help a lot i just wish the videos where longer, by the way your a great teacher . . . im so anxious to learn much more . . . meanwhile im only on your first . . . lol. . . . i know patience right?
> im picking it up though. . . . i love the congas.i actually learned the timbales which i think is a beautiful instument . . . thanx for taking the time to teach new comers like me. . . . QUE VIVA LA SALSA. . . .

> prpapito3000 (reply): Thas great to hear . . . what a good christmas gift! lol, wish i got that for christmas . . . but yea just work on this and the other videos will help you to keep developing more and more!

> ben5r: silly hat!!

> prpapito3000 (reply): lmao. . . . HATER! . . .
> (comments posted on prpapito3000 2007a)

The Sanchez video and the Torres video were both posted in 2007, relatively early in YouTube's history. This partly accounts for their placement in the search rankings; they were among the first conga lessons uploaded to YouTube, and once they had attracted a tipping-point number of views, they were positioned to maintain their view-count dominance for years to come. Neither video was posted as a teaser for another online business or for private lessons. While the Sanchez video appears to be excerpted from a commercially available DVD, that DVD is not linked from the YouTube page or cited in detail (indeed, the title is misspelled—probably a typo, although sometimes an indicator of deliberate evasion of copyright enforcement). Torres just seems like a friendly teenager who's messing around with YouTube and imagining what he could contribute to the available content.

These YouTube videos and their associated comment threads offer two contrasting examples of A2A pedagogy. The first uploader, Gordanius, has posted a professional musician's lesson as a public service; in the comment thread, viewers demonstrate their own expertise by offering criticism, defenses, and their own advice to learners. The second uploader, prpapito3000, is an amateur himself, but he certainly knows enough to offer useful lessons to beginners. Moreover, he's willing to interact with his viewers online, answering questions, offering encouragement, or brushing off typical YouTube slagging about his wardrobe with a format-idiomatic "lmao" ("laughing my ass off"). His youth and nonprofessionalism actually make him appealing. After all, if I'm watching this video, I'm probably playing congas in my living room, too. And if I'm still just learning how to produce an open tone, do I really need to get a lesson from a venerable professional performer? It's also quite likely that I'm only aspiring to be an amateur player myself, in which case maybe Nate Torres is a better role model for me than Poncho Sanchez. He's certainly more accessible:

> amigosjh: Could someone tell me how to correct the open tone?. Mine sounds weak.

> prpapito3000 (reply): You should make sure to keep your fingers close together and aline your knuckle line with the edge of the rim . . . take a look at my other video for "basic tones" and that should help as well!

> aureomarinho: Hi there Nate!
> Greetings from Brazil!
> Thanks for the unique videos! Help me a lot to correct some mistakes! Keep on doing this! Latin music congas is everything! The soul, the bones of the rhythm! Thnx

> prpapito3000 (reply): thas right brother! . . . keep it up and hope to be of help to you
> (comments posted on prpapito3000 2008a)

I contacted Nate Torres through his YouTube channel in December 2009 and arranged a phone interview. At that time he was 20 years old and a college junior, although his YouTube profile said he was several years older. (Like many YouTube users, when Torres first created a user account he had to misrepresent his age in order to meet YouTube's minimum-age requirement.) Torres is of Puerto Rican descent—as the screenname prpapito3000 implies to some of his audience—and he comes from a musical family:

> My grandfather, he was a part of a lot of different bands, and he was in the whole nightclub scene for a long time in Puerto Rico. And then my father moved to New York, in his teens. And he was playing in New York as well, doing the whole nightclub thing, and going from place to place playing gigs.[2]

The family moved from New York to Virginia when Torres was a child (though he still lists the Bronx as his hometown on his YouTube profile). He asked for a set of bongos for his ninth birthday, but was given conga drums instead. Torres learned some conga technique informally from his father but received formal instruction almost entirely from video lessons or filmed performances: "I would go to the music store and buy video lessons and learn that way. . . . I guess I would have liked to have personal lessons. I always lived around this area [in Virginia], so there weren't many options available for this type of instrument."

As with the teachers discussed in Chapter Five, Torres's own learning experience informed his approach to making YouTube lessons. Years of learning from video lessons made him attentive to useful visual cues and led him to find ways to encourage students to self-assess and self-correct. Moreover, the fact that he had to rely on video lessons for formal instruction—due to the dearth of conga instructors in his area—led Torres to think about the plight of aspiring musicians who couldn't even find or afford videos.

KM: So what originally inspired you to make these video lessons and put them on YouTube?

NATE TORRES: I guess the fact that when I looked around, there wasn't much there at the time. . . . And I guess at the time I was looking for videos, there really wasn't much that went from the beginner to intermediate level. I kind of wanted to just share with whoever's willing, or whoever wanted it, the information that I had gained, or the knowledge that I had gained. Because there's a lot of people—you know, YouTube is a free source. I've spent a lot of money on videos. Videos aren't cheap, they were 30 or 40 dollars apiece at the time that I got them. You know, because there wasn't YouTube or all these other websites where you can get stuff like that. So I figured, there's a lot of people out there that can't afford it or aren't able to spend money on it, so I wanted to try and help those people out. . . . Surprisingly, I got a lot of views and a lot of comments. And I guess I didn't know the extent that people would really appreciate it until I started getting a lot of messages and comments.

One might think that Torres's long experience with professionally produced video lessons would discourage him from creating his own lessons, given his limited credentials. Instead, it seems that he became increasingly aware of his privileged position: He had familial support, could claim this musical tradition as part of his cultural heritage, and could afford professional video lessons. When he saw that few conga lessons were part of the growing YouTube archive, he decided to fill the gap himself—in the spirit of what Robert Stebbins calls "contributive participation" (1992:121).

Nate Torres's YouTube channel includes only nine lesson videos—all uploaded in 2007 and 2008—plus a handful of videos of him playing songs with family members. At the time of our interview he had not posted a new lesson video in eighteen months, but he continued to receive regular comments and private messages from viewers. Commenters often thank him profusely and tell him that he is the clearest teacher they have ever encountered. Many ask him to create lessons teaching the basic rhythms for particular Latin Caribbean genres, including merengue, bomba, plena, and reggaeton. Commenters post in both English and Spanish; many mention that they live outside the United States (for instance, in France, Mexico, the Netherlands, England, Malaysia, Canada, Colombia, and Brazil). Some commenters tell Torres that the lessons are improving not only their technical abilities on the conga, but their listening or dancing skills:

> juhanleemet: Good stuff! I'm learning a lot. One "interesting" thing I noticed is that the "hand" on the downbeats 1 and 3 produce a kind of "pulse". . . . Is that the right way to think of it? Like a "pulse" on the downbeat, and slap and open tones on the backbeat? BTW, I can now recognize the tumbao moderno, as in a Gloria Estefan tune I heard recently: There it was! Large as life!
> (comment posted on prpapito3000 2007b)

> clc38: Gracias amigo! I am a salsero who has been meaning to learn the basic salsa percussion for a long time . . . it will really help my dancing and understanding of the musica.
> (comment posted on prpapito3000 2007c)

Once in a while commenters post criticism of Torres's playing or his pedagogy. Some write as though giving friendly feedback to a student; others admonish Torres for perceived offenses to tradition. Torres always responds.

> arouttenberg: Nate, you certainly have a gift for teaching. Great videos! The only comment I have is to be careful with the technique on slapping. It's ok to "cup" your hand when you first start to play . . . but you gradually should be able to produce a slap without having to cup your hand.

> prpapito3000 (reply): uhh . . . I would have to disagree with you on that . . . yes, you shouldnt cup your hand excessively however the correct technique is cupping your hand to produce a clean slap and it really doesnt produce any type of "bad callus" . . . thanks tho
> (comments posted on prpapito3000 2007b)

> cisgom70tube: very important is the BASS note you are Missing in the second note of the clave!!!. . . . if you refer to guaguancó Cubano tradicional, then you need to teach it traditionally and that's not the way. Learn the basics first

> prpapito3000 (reply): I never referred to it as "tradicional" and thank you I do know the basics. I definitely am not the best player, however, I do know what I

am talking about. And if you feel so strongly about my videos then you can easily post your own to show the "correct" way that you so much proclaim to know. Thank you. . . . also, if you listen I said this is the EASIEST way to play it not the "traditional" and if you look at the next video (#9), then you will see that there I show it the TRADITIONAL way which is a little harder!

(comments posted on prpapito3000 2008b)

However, critical comments are generally few and far between, and when they do appear other commenters often come to Torres's defense. As one fan wrote after Torres engaged in an unusually lengthy exchange:

hey man. fuck those haters!! they just dont have the confidence to make their own vids. writing negative Comments is soo ridiculous!! you all have to be happy to get so nice lessons from this cool guy!! he took his time for you fuckers to make you a better conga players or to show basic skills for new players. teacher want money for it and their dvds. he not!!! just shut your mouth

(comment posted on prpapito3000 2007b)

Torres's amateur status seems to provide some protection from critical comments of the kind that appeared on the Poncho Sanchez video. The style and content of Torres's videos make it clear that he is an intermediate student intent on passing his own knowledge forward, rather than a self-appointed expert vying for professional recognition. Of course, the fact that he is of Puerto Rican descent also affords Torres credibility with some viewers, including both Latinos and non-Latinos who believe in a blood-based "Latin touch."[3] Commenters with the knowledge required to question Torres's expertise generally offer constructive criticism; they seem glad that this Puerto Rican teenager is taking an interest in traditional conga technique at all.

Still, some commenters do imply that Nate Torres shouldn't be rubbing shoulders with Poncho Sanchez or other recognized professionals as a conga teacher. Now that there are plenty of professional musicians and teachers posting conga lessons on YouTube, why should anyone bother with lessons from prpapito3000? I suggested one answer already: For brand-new beginners who need instruction on producing basic tones on the drum, Nate Torres might be as good a teacher as anyone. His lessons might also seem more accessible and appealing to some students precisely *because* they are presented by a teenage amateur. To me, though, the most important reason to visit Nate Torres's YouTube channel is that he teaches beginning conga students much more than basic tones and the *tumbao* pattern. He gives them a glimpse of conga drumming as an everyday musical practice in the life of a Virginia college student of Puerto Rican descent rather than as an exotic bit of "Latin flavor" or something you'd find onstage in a salsa club. For example, viewers who visit prpapito3000's YouTube channel will see not only his basic conga lessons but several videos of him playing congas at his church—probably not what our hypothetical new salsa fan expected when she or he went looking for basic conga lessons on YouTube. Torres also uploaded one video in which he plays congas to accompany a performance by his Latino fraternity's step team.

Torres's church videos led some viewers of the regular lesson videos to post questions about his religious affiliation:

> guapo4u: nate i watched your How to NOT play Congas or Drums video and saw the sign on the wall are you jehova's witness? just wanna know bro.

> prpapito3000 (reply): lol uhh, no im not a Jehovas Witness . . . I am a protestant christian (pretty much meaning im not catholic) lol . . . but thats cool, I dont mind the questions
> (comments posted on prpapito3000 2007c)

For other viewers, the church videos functioned as the entry point to the lesson videos:

> KM: I saw a bunch of the channel comments seemed like they came from church connections, or they were from people who would use religious language. And I was curious if you knew those people from elsewhere or how they knew that you were involved with the church?
>
> NATE TORRES: On my channel, I have other videos that are of me playing. And it's pretty obvious, if you see those videos, that it's done at a church. It would be, like, on the altar. So I think once people see that, then they assume that. And then they'll leave comments saying "God bless" or "Thank you, God has given you talent" or things like that. . . . A lot of those videos are tagged using some church language, I guess you can call it. In Spanish. And, you know, we have words for hymns and church song types, so if you type that in, then they'll definitely come up.

Because he posted videos of his church performances, Torres often received messages from people who wanted to include conga drumming in their worship music:

> One recently was from somebody who told me that they were Jamaican, and that they were playing the congas in the church, and that they were trying to incorporate it more. I've also gotten messages from African American churches, or people that are musicians in those types of churches that typically play contemporary music. And they're saying that they want to incorporate this instrument in their performances.

As Nate Torres's YouTube channel demonstrates, in A2A pedagogy both teachers and students often let signs of their broader social lives mingle with their lessons or accounts of their learning progress. This is partly due to the nature of amateurism: Amateurs are often eager to share their leisure pursuits with friends, and they may see no reason to establish the boundaries that delimit professional teacher–student relationships. Moreover, A2A operates through online platforms with built-in social networking features: detailed user profiles, opportunities to link to "friends," invitations to post response videos, "favorites" lists, and so forth. As far as playing congas is concerned, Nate Torres uses the same techniques to bridge virtual and visceral experience as the professional teachers discussed in Chapter Five: He finds ways to communicate the embodied experience of playing the drums. But his YouTube channel also offers viewers other kinds of bridges, including access to aspects of his family life, cultural background, religious experience, and social world at college.

HOW TO PLAY "PIANO MAN"

> Hey John, I downloaded some of your videos just in case youtube goes bananas on you. I do hope your channel stays intact though, because you have really fostered my love for the piano and everyone deserves to have you as a teacher. (HZTBB, comment posted on pianojohn113 2007b)

I first heard of pianojohn113 when I was reading the comment thread for WNYC radio's "Smackdown" debate about online music lessons (see Chapter Five). One listener wrote, "Online the best tutorials have a camera well positioned to see the instrument exactly as you the player can. That is a nice answer to an old problem for in-person lesson format! and a million thanks to pianojohn113!" ("Phillip from bklyn," comment posted on WNYC 2010). A Google search led me to pianojohn113's YouTube channel, where I found about 150 piano-tutorial videos, nearly all of them demonstrating how to play songs by the Beatles and Billy Joel. Most of the videos had been viewed over 10,000 times; 22 had surpassed 100,000 views, with the most popular being the lessons for the intro sections of the Beatles' "Let it Be" and Billy Joel's "Piano Man" (each garnering over 700,000 views between 2007 and 2011). By March 2011, pianojohn113's total viewcount was nearing 11 million, and almost 20,000 people had signed on as channel subscribers.

In all of the tutorial videos, the camera shows an overhead view of pianojohn113's hands on a piano keyboard. Occasionally the top of his head dips into view at the bottom of the screen, revealing graying hair, but otherwise students only see their teacher from wrists to fingertips. As he talks students through the lessons, text boxes occasionally pop up over the keyboard showing the names of the notes he is playing. Most of the videos are about 10 minutes long, maxing out YouTube's file-size restriction for uploads; several of the song tutorials are broken up into a few separate videos to get around this restriction.[4] Here is what pianojohn113 says in the first two and a half minutes of the lesson "Billy Joel #1–How to Play Piano Man Intro":

> Starts off with the left hand playing a D-minor 7 [*plays chord, text box shows note names over appropriate keys*] and the right hand is simply going to be an A-minor 7 run-up [*plays arpeggio*], A C E G, but starting with the A-flat [*plays phrase twice*]. Very easy [*plays phrase again*]. Now from the G you're going to have to go down chromatically to the C, and you can do it however feels comfortable for you. It might take practice if you're not used to doing chromatic scales, which is every note, all the way down [*plays chromatic run*]. I like to end—[*plays start of intro again*] I like to play the G with my third finger [*plays G*], because it's easier for me to work it down to the C [*plays chromatic run*]. Any way you feel comfortable doing it, do it that way [*plays whole phrase*]. Next part, left hand [*plays chord*], D minor, but diminished [*plays chord*]. D F A-flat B [*plays*]. And you're going to do a few rolls in the right hand. The first one is B-flat B C [*repeats roll several times*]. Billy plays it, Billy Joel plays it twice [*demonstrates*], and sometimes he plays it three times [*demonstrates*]. You could play it six if you wanted, it really doesn't matter. [*Plays roll several times, adding end of phrase.*] But if you want to play it like he does, play it two or three times. I like three. [*Teaches several more rolls.*] That's that part. Now the rest of the intro goes like this. If you have—I'm going to play this as if you *don't* have a harmonica. If you do have a harmonica it's easier, because you're just going to play the chords. (transcribed from pianojohn113 2007a; see ▶ 6.3)

The 819 comments on this video (as of April 2011) show that pianojohn113 has managed to hit a pedagogical sweet spot. He wins thanks and praise from both absolute beginners and more experienced players, from people who play by ear and people who read notation. The video format makes it possible for beginners to pause the video every three seconds to laboriously place each finger in the right position on the keyboard; more advanced pianists can instead play along with the lesson at it proceeds, taking advantage of the frequent pauses, short repetitions, and bits of music-theory scaffolding that help them quickly absorb and remember the material.

Many commenters write that they have purchased sheet-music versions of the song and have been disappointed by the arrangements. They want to play exactly what Billy Joel plays, not a simplified version with left-hand chords and right-hand melody doubling the vocal line.

> lovalouie: You are friggin' awesome. Over the years, I have purchased hundreds of dollars worth of books and sheet music, only to be disappointed by the fact that only a few of them include complete intro and solos. I should have left "Billy

Joel Complete: Volume 1" at the store, and sent you the $27.95 instead! Thank you for taking the time to post these tutorials.

Others explain their own relationship to the song, seek encouragement, or ask for technical tips:

Trudyisluved: I love your videos!! This was my dads favorite song and I wanted to learn it just for him! I'm 13 and have only been with the piano for 11 months. Do you think I could learn this song by memory and as fast as you go by May?

pianojohn113 (reply): Yes, if you're good with octaves and are familiar with the chords and progressions, I think you can learn it. I don't know what your skill level is, but the verses are all exactly the same, so if you learn one, you'll know them all. so give it a shot!

elrais2: I know you answer questions a lot so . . .
Will it sound fine just playing octaves in the left hand instead of using three fingers?
I think I can keep the rhthym, its just hard to do so many fingers. thanks in advance

pianojohn113 (reply): That's a really good question. I think it will sound fine, although the middle note in the LH is the root of the chord on most of them here. It might change the flavor of the chord just a bit without it, but I understand the difficulty in using so many fingers. it can be a stretch for many people. So yes, feel free to just play the octaves!

The thoughtful, supportive, and measured tone of these replies is typical of pianojohn113's teaching persona; he never seems to lose patience with pianists who find a three-finger chord challenging or who don't feel qualified to decide for themselves whether a variation "sounds fine." He replies to nearly every comment posted on his videos, including the rare few that express criticism:

sophiesmurph: i dont like the fact that you put in some of your own notes it kinda makes it your version of piano man not feckin billy joels!

pianojohn113 (reply): Nothing should ever be copied note-for-note. even Billy Joel adds his own fills and phrases each time he plays it. Listen to his live versions, he never plays it the same way twice.

sophiesmurph: yeah but you said quote "Not "easy piano," but it's the way Billy Joel plays it." [This description is in the video info section.]

pianojohn113 (reply): Right, exactly. The chords are Billy Joel's. The structure and fills are Billy Joel's. The arrangement is Billy Joel's. The interpretation is mine. Way different from the "easy piano" versions that are out there, which is why I mentioned it.

When I was reading this comment thread, sophiesmurph's comments had been hidden because other viewers had flagged them as negative content (using YouTube's "thumbs-down" button); I had to click "show" to read them.

Pianojohn113 does not create or distribute sheet music, sell his lessons, or bring in money on the side by signing up to have YouTube display ads with his videos. His tutorials are strictly an amateur-to-amateur affair. The last entry in the FAQ included in his channel profile explains why:

Q: Do you have any DVDs of your video lessons available for purchase?
A: No, I do everything for free right here. I really believe that music should be shared and learned by everyone who wants to play. If my videos add to your musical enjoyment or help you to become a better musician, then it's worth it for me. (pianojohn113 2007b)

Commenters often thank him profusely for providing the tutorials for free, while also clamoring for more lessons. As David Taub and Nate Brown discovered when they posted guitar and drum lessons to YouTube, many people seek out such lessons because they want to learn particular songs (see Chapter Five). Like Taub and Brown, pianojohn113 constantly receives specific song requests:

jleperchan: i love ur tutorials and wonder if u could teach how to play "love" and "jealous guy" by john lennon. I know ur probably not big punk fan but i was wondering if u wuld be interested in teaching how to play "Rock the Casbah" by the clash. Also do u have any interest in Lou Reed? thank you much appreciated and keep up the great work

YoungMerc08: The One! by elton! please!!!! especially the solos! and you know all those litle licks and tricks that elton does especially on his late concerts? if you can do those, would be great if you can do a video teaching just how to do those . . .
(comments posted on pianojohn113 2007b)

Most requests are framed graciously. When some commenters were less polite, a friend leaped to pianojohn113's defense:

Beatlefanatic9: You should learn how to play Free As A Bird, Also you should have a video where your playing the whole Abbey Road Medley.

fdannen: Memo to Beatlefanatic: You should learn how to say thank you.

enolaniaga: John, Watching the Wheels by Lennon would make a great tutorial, ecspecially for begginer/intermediate piano students. It sounds great on just piano and is all white key playing.

shanebro2: Response to enolaniaga: I asked John to do "Watching the Wheels" already many months ago and he hasn't done it.

fdannen: Reply to Shanebro: John and I are friends, and I've asked him many times why he tolerates unschooled, ill-mannered ingrates like you. You complain that you asked him for a tutorial and "he hasn't done it." Who the hell are you to give him orders? Are you paying him a salary?
(comments posted on pianojohn113 2007b)

The negative tone of this exchange is quite unusual for pianojohn113's channel. Commenter fdannen uses strong language to defend the civility that is the norm here, in stark contrast to much of the discourse on YouTube. As Nate Torres's experience also demonstrated, the fact that A2A teachers don't charge for their lessons tends to shield them from "flaming" or even from rudely phrased lesson requests. Most students seem to feel beholden to their volunteer teachers, and they repay them by posting expressions of gratitude that set the tone for other commenters.

I contacted pianojohn113 through his YouTube account in the spring of 2010 and arranged an email interview. John Dean is originally from the Kansas City area but has spent most of his life in Southern California. He was 46 years old at the time of our interview and had been playing piano since he was three. Dean is a radio producer and co-host for a nationally syndicated financial talk show; he also owns an audio-production company. Dean's YouTube profile includes his age and some information about his musical background, but it does not mention his job. Most of his viewers probably have no idea that he works in the media and can be heard on a daily radio show. He told me that he originally included his email address and work website in his profile, but removed the information after an overzealous female fan tracked him down and "showed up at my front gate, wanting to know if this was Piano John's house." Dean has only met one other student in person—"just someone who heard me playing at a hotel party and thought I sounded familiar. . . . He just wished me well and said thanks." However, he obviously values his online interaction with viewers, since he makes a point of replying to every personal YouTube message and most public video comments that he receives. Dean derives great satisfaction from his students' progress: "The best comments and messages I get are from people who say they gave up piano when they were young, and then they saw my videos and were inspired to take it up again. Some of them even post videos of what they've learned, performances, both public and private. . . . It's the reason I continue to put the tutorials up."[5]

Dean believes that video lessons on YouTube are "much more beneficial" than most face-to-face lessons—although also more difficult for the teacher, because "that person must make a video lesson that's suitable for a wide variety of abilities and musical tastes, without actually knowing who their students are." He outlined the same benefits highlighted by the teachers discussed in Chapter Five—affordability, repeatability, convenience—but he particularly emphasized student autonomy. Free online lessons liberate students from "the pressure to get your money's worth" and "the pressure of having to 'perform' in front of someone who is judging them and their talents"; they also let students choose which songs they want to learn "and then decide for themselves if the song is too easy, or too difficult, or just right." Dean's own childhood experience with piano lessons informs his convictions about their shortcomings:

> Here's what most young kids have to go through, typically: They are assigned a song to learn, one they don't care much about, or one that takes a long time to master. They are required to practice X number of times daily, because lessons aren't free and practice yields results. Come home from school, practice the piano, just like homework. Not fun. They usually struggle through a song for a few weeks, stumbling around until they finally get to a point where they can kind of play it most of the way through without too many mistakes. At that point, the teacher marks the song off in the book with a check, or puts a sticker on it, then turns the page, and the student starts the process all over again. The poor kid is always struggling, playing songs they may not want to play in the first place, never really learning a song that they like. The constant fumbling around just kills all the fun. As a result, they wind up quitting. That was my experience, in a nutshell (although I didn't give up piano). I wanted to go back and enjoy playing the songs the way I did when I was 3 years old: hear a song I like, mess around with it on my own time, try to pick out the melody, and just have fun with the whole thing. That's the method I try to use on my video lessons, and the message I try to convey to others who have had the same experience I did.

Essentially, Dean had to relearn how to play for the pleasure of playing; he constantly encourages his students to do the same thing. He has experience as a paid performer—in his 20s he often played the "dinner music" repertoire at events and private parties—but he is now a passionate advocate for amateurism. This message seems to get through to his viewers. As one wrote in a channel comment:

> Hey John, you provide a great service for all of us who play music for reason other than financial gain. I have done that for many years and cheated myself of

personally enjoying music [. . . .] I have finally found a reason to continue playing music other than making a living playing songs I can't stand. The Beatles
were what made me want to play and seeing your work dedicated to them is
wonderful. (harmonyd1, comment posted on pianojohn113 2007b)

Dean also encourages amateur performance by lavishing praise on students
who post their own YouTube videos. For instance, when TheYellowSubma-
rine13 asked him to check out her rendition of John Lennon's "Real Love,"
pianojohn113 posted this comment:

This is fantastic! What a pleasure to listen to your performance. The piano is
great, and your singing is equally good. I had no idea you could sing that well!
This could be a "signature" song for you, one you'll be playing for many years.
Awesome job, really, really, REALLY well done! Thanks for posting it. –John
(comment posted on TheYellowSubmarine13 2010)

According to her YouTube profile, this performer is a 14-year-old girl who
is obsessed with Jesus and the Beatles. (Her face does not appear in her videos,
perhaps as a nod to privacy concerns.) She explains in her profile that her
video uploads are relatively infrequent

because I have to (1) stop being lazy and convince myself to learn a new song (2)
actually learn the song (3) practice it enough so that I can play it perfectly (4)
convince myself to stop being lazy and make a video (5) actually make the video
and make sure it's perfect, and, occasionally, there's a 6th one: learning all the
lyrics if I have decided to actually sing in the vid! (TheYellowSubmarine13 2009)

This self-assessment offers an excellent example of the autonomy that Dean
attributes to online music students. It isn't that they don't practice but,
rather, that their practice doesn't feel like drudgery. TheYellowSubmarine13
uses pianojohn113's lessons to help her develop and perform a deeper relationship with her favorite music. Her YouTube videos aren't showy, and she
gives no indication of wanting to become a professional musician; her profile
states that she aspires to become a photographer for *National Geographic*.
Instead, it seems that she makes the videos because they motivate her to
improve her musicianship, and so that she can publicly document her devotion to the Beatles and share it with other fans (cf. Jenkins 2006b; Ito, Baumer et al. 2010a).

In March of 2010, pianojohn113's tranquil island of classic rock piano lessons
and civil discourse was disturbed by a copyright infringement notice. The video
alerting viewers to this turn of events stands out starkly among the tutorials
(see Figure 6.3). White text scrolls against a black background, communicating
the bad news and pianojohn113's tactical response:

Blackbird - Beatles (piano) DEMO
5,274 views
2 months ago

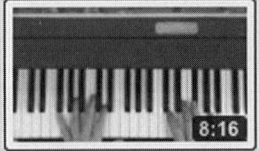

Lucy In The Sky piano tutorial
7,546 views
2 months ago

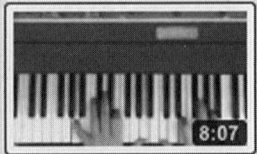

All You Need Is Love piano tutorial
6,732 views
2 months ago

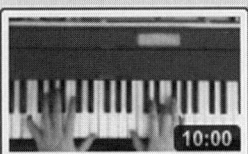

Billy Joel's "Lullabye" tutorial Part 2
3,664 views
3 months ago

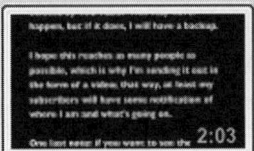

A Pianojohn113 Bulletin!
4,812 views
3 months ago

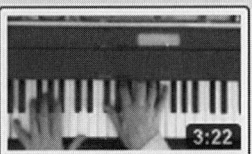

Billy Joel's Lullabye INTRO tutorial
5,928 views
4 months ago

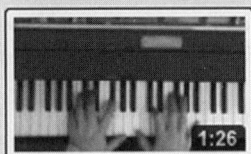

Drive My Car - Brief demo
8,196 views
4 months ago

Good Day Sunshine piano tutorial - Part 2
1,729 views
4 months ago

Good Day Sunshine piano tutorial - Part 1
5,819 views
4 months ago

Figure 6.3
Pianojohn113 upload index, including copyright bulletin (pianojohn113 2007b). (*Screenshot by the author*)

An important notice from pianojohn113:

Recently, my two tutorial videos for the Eagles' "Desperado" were removed by YouTube after a copyright complaint was filed by Cass County Music, Inc.

They have been searching through YouTube for any and all Eagles video covers, tutorials, live performances, etc. and demanding that they be removed immediately. This is what happened to my two videos.

YouTube has a policy that dictates, among other things, that too many "copyright violations" will result in that person's channel being closed down. The limit, I have been told, is anywhere between 3 and 5 videos.

I will not argue against the terms of the copyright issues. If Cass music believes it is a violation to post self-made tutorial videos of Eagles songs, then they have the right to demand the removal of the video.

What I do NOT want to have happen is the complete cancellation of my pianojohn113 YouTube account [. . . .] So, I have set up an ALTERNATE pianojohn account, which will eventually contain all of my current videos, in the unlikely event that my account is shut down. It will be simply this: pianojohn113a [. . . .] Once I have re-uploaded all of my current videos, it will mirror exactly the pianojohn113 channel. That way, I can keep the videos posted and not have to change my YouTube identity too much [. . . .]

I hope this reaches as many people as possible, which is why I'm sending it out in the form of a video; that way, at least my subscribers will have some notification of where I am and what's going on.

One last note: If you want to see the "Desperado" tutorial video, just send me a message, and I will direct you to where it is.

Thanks, all. Keep on rockin' the piano. (pianojohn113 2010)

John Dean believes his tutorials are covered by the "fair use" provision in U.S. copyright law (U.S. Code Title 17, Section 107), but he was not prepared to challenge YouTube's decision or Cass Music's allegation of infringement. As he told me, "If I really wanted to fight Cass Music, I probably could have done so and won, but at what cost? . . . They probably figured nobody would have the time or inclination to fight it, and they were right, at least from where I stand."

The bulletin video had been viewed 8,444 times by April 2011, and it generated a flood of supportive comments:

dormanbenton: I just wanted to say we'll follow you wherever you go Pianojohn. Even if we can't find you instantly . . . we will. There's only one person putting these tutorials out. You! You are the best and most viewed. Those copywrite issues are B.S. and obviously they don't care if it's actually helping the artist. We wouldn't be learning these songs if we didn't love and admire the band/artist.

PedroElVirtu: I'm a 50 years old guy who never took a piano or keyboard in my hands until i saw your channel about almost 2 years ago and now i can play around 20 songs almost perfect thanks to you, totally free without no demands or conditions from your part. If youtube wants to close your channel there will be nothing you can do to stop that, and believe me, if they close this channel they will close the new one too [. . . .] i say you should open an account in facebook. Lets keep the tutorials over there in a new account that will not be part of your real one where you have your friends and family, but remember that in this almost 15.000 subscriber there are lots of people that have you as a family too and in my case, i don't wanna lose you my friend. I hope you get what i'm trying to say in my limited english and you let us know how it's gonna be the solution of this.

(comments posted on pianojohn113 2010)

Dean told me that despite his students' promises to follow him any-where, "I never really considered leaving YouTube, because right now it's the most well-known and easily accessible way for people to view videos online." Moreover, his videos are embedded in hundreds of other webpages using YouTube's embed code, so if he lost the channel or left YouTube "all of those other sites would be affected as well." Some commenters suggested that Dean fight the takedown order; others proposed some kind of commu-nity protest movement, like a petition or boycott. Many more seemed to assume that total account deletion was inevitable and imminent. Shortly after this video was posted, many viewers posted comments on the main channel page thanking Dean for all of his work, essentially saying goodbye in advance.

In many respects this copyright episode echoes David Taub's experience with posting guitar song lessons on YouTube (see Chapter Five). Both teachers believed their videos were covered under "fair use," but neither was prepared to fight a legal battle or to rock the boat with YouTube, where they had already built a relationship with thousands of channel subscribers. However, David Taub's position as a professional teacher with a web-based business gave him some options: He could use YouTube to attract students to a separate website, and he could continue to offer song lessons on DVD by paying for the rights with revenue generated by his teaching. John Dean is much more dependent on YouTube, in part because he has no interest in going pro; he is committed to A2A music lessons on principle.

Dean told me that his response strategy had been to "comply completely," because "I have way too much time and effort devoted to my channel at this point to jeopardize the account." However, his channel includes some gestures of resistance, including the information in his copyright bulletin video and this note in his profile:

> Just a reminder: In the event that my account is deleted by YouTube, see my backup channel: pianojohn113a. My current videos are slowly being added there [. . . .] A fellow YouTube user has generously transcribed my version of the [un-named] song that was deleted from my channel by the copyright owners (rhymes with "Hesperado," by a group that rhymes with "The Heagles") and has made it free for everyone to download. (pianojohn113 2007b)

A year after the "Desperado" snafu, Dean's YouTube account had not been deleted, but he had received some new takedown orders. On his channel profile page he reported that he had been forced to take down his tutorial for Elton John's "Candle in the Wind" because it included a brief demonstration at the beginning: "I was told in very harsh terms that the Music Publishing Rights Collecting Society, whatever that is, has complained that my tutorial has cost them untold millions in revenue, and another 'strike' against my account could be forthcoming. . . . In the new video, the tutorial section is still there, but the demo portion was edited out. We'll see if that placates the critics" (pianojohn113 2007b, 2011 update). Dean offered further context in an email:

> You Tube (and music publishers in general) have seemed to back off somewhat on going after tutorial videos, although they have flagged a couple of my quick demonstrations of the songs where I play several bars of whichever song I'm teaching. They have not asked that I take them down this time; rather, they say that they would like to reserve the right to put advertising on my performance videos if or when they decide they'd like to. I suppose they've determined that the amateur reproductions of portions of their songs is allowable, at least when connected to a tutorial. I'm not sure if this is just temporary amnesty, or if they'll go full speed ahead again in the future, but I am still proceeding cautiously.[6]

Meanwhile, in various comment threads, pianojohn113's students continue to discuss fair use doctrine, whether blame should fall on the Eagles or on evil industry executives, and their gratitude for Dean's unpaid work. A2A may be teaching them as much about copyright and competing definitions of musical labor as it is about classic rock.

CYBERSHALA: THE ASHTANGISPHERE NEVER SLEEPS

> I'm not really sure what the Cybershala is but I've heard it referred to more and more lately. It seems to be an online community not located at any one site or of a fixed membership. It seems to be made up of blogs, comment threads and forums, corners of Facebook, YouTube, chatrooms, and Skype connections. Anywhere where one's practice can be posted, discussed, commented on . . . (Grimmly 2010: part 17; see Figure 6.4)

On a break from writing about pianojohn113, I logged into my Facebook account. A friend had posted a comment on one of my status updates, asking about my backbends.

> Jessica: how are your back bends coming? are you dropping back unassisted now? (I seem to recall you saying you were close . . .)

> Kiri: not yet, still close—the 2nd series backbends are helping. Trying to be nice to my tweaky left shoulder.

> Jessica: They've made a big difference for me, too. I used to hate backbending and now I look so forward to it every day! How far are you up to in 2nd now?

Figure 6.4
Grimmly working on his drop-backs on YouTube (grimmly2007 2010a). (*Screenshot by the author*)

Kiri: Kapotasana, but with a lot of help from the wall. [. . .]

Jessica: my kapotasana is getting easier, partly thanks to this video. I don't know why, but i'd never put together that pushing my hips forward would make it easier. Also the relaxed upper body part helped me feel it differently. His drop-back vid was helpful too. http://www.youtube.com/watch?v=F_GXujI-fqA
(July 29, 2010)

My friend and I both practice ashtanga yoga, a method codified and popularized in the twentieth century by Sri K. Pattabhi Jois (1915–2009) through his Ashtanga Yoga Institute in Mysore, India. Ashtanga yoga features many poses (*asana*) familiar from other yoga styles. It is a "flowing" yoga style, in which most asana are held for only five breaths; fluid transitions between asana are an integral part of the practice. The simplest explanation of what makes ashtanga distinctive is that practitioners always move through the asana in a particular order, matching their movements to a particular breath pattern. However, to experienced practitioners this description makes about as much sense as saying that a clarinetist produces sound by matching her musical output to a particular breath pattern. Though it may be observably true, it doesn't convey the crucial fact that breath is *required* to produce and sustain the music/the movement. The breath technique (*ujaii*) is audible, sometimes described as "ocean-sounding." While few practitioners would describe what they do as "musical," their breath practice undoubtedly qualifies as "humanly organized sound" (Blacking 1973). Practitioners pace themselves by controlling and counting their own breaths, and teachers use the sound and rhythm of students' breathing as a diagnostic tool: Is the student comfortable or struggling, focused or just going through the motions?

The prescribed sequence of asana, along with their associated breath counts and gazing points, is broken down into a structured curriculum: fundamentals, primary series, second series, and so on. Students learn each series through repetition and can then practice independently. At "Mysore-style" ashtanga classes, the teacher does not instruct the whole group and rarely demonstrates asana. Instead, she or he moves through the room giving physical adjustments and working with students one-on-one as they proceed through the series. (Imagine a room full of dancers all performing the same choreography, but at different paces and with staggered start times.) The teacher decides when a student is ready to add a new asana or begin a new series. At the time of our Facebook exchange my friend and I were both partway through learning the second series, 3,000 miles apart and working with different teachers.

That week I had been thinking about adding a dance case study to this chapter. Every ethnomusicologist at some point finds herself explaining that in many cultures music and dance aren't separable categories; moreover, web-based dance transmission resonates with the themes of virtual and visceral embodiment that run through this book. Dancers play their bodies like instruments; musicians master a stylized dance with their instruments; and technical interfaces "do a sort of social/emotional 'dance' with the end user" (Isbister and Höök 2007:4). Dancers who learn to perform particular styles or choreographed repertoires reshape their bodies through repetition, as instrumentalists do. They might not have to adapt to the physical qualities of a particular material interface—although some do, for dance styles that employ props or elaborate costuming—but they continually experiment with the potential and limits of their individual bodies (Hamera 2007).

Many dancers now use video-recording technology to document choreography, distribute performances, review lessons, or learn without a physically present teacher. Like online music students, they must grapple with the challenge of translating a two-dimensional video image into three-dimensional embodied practice. For instance, Tomie Hahn discusses her teachers' ambivalence about the use of video recordings in the transmission of Japanese *nihon buyo* dance. Video is considered useful as documentation or as a reference point for distant students, but "dance is not dance without presence. . . . While videotapes supply visual images, personal contact provides emotion and individual nuance." Hahn's teachers "acknowledge the practical aspects of video technology, [but] they are very clear, even passionate, about its limitations" (2007:142).

Until my friend posted the link to the backbending video, it hadn't occurred to me that all of these affinities between music and dance also extend to yoga, or that web-based resources might be playing a significant role in yoga transmission. I had been practicing yoga for the preceding eight years in large part because it took me away from my computer. But I clicked on the backbending-video link and watched a man work his way into kapotasana. He provided voiceover commentary on the process, supplemented with arrow-shaped text boxes drawing attention to "hips forward, heart open, arms and chest as soft as possible" (lrockwood 2006; see Figure 6.5). In YouTube's "related videos" sidebar, dozens of yoga teachers and students offered videos tagged with "kapotasana"—some providing instruction, many others simply documenting their progress. Several people had uploaded the videos to YouTube so that they could embed them in yoga blog entries, where they discussed their struggles with particular asana in extraordinary detail. Consider this blog post by "The Reluctant Ashtangi":

> Holy smokes!!! I almost stood up from backbend today!
>
> I was in early, the assistant was there and no one else. It was like having the Shala to myself! I settled into my favourite spot and enjoyed a very inward-directed practice. The room was completely silent except for the sound of my Ujjayi and I moved through my Primary steadily. [. . .]
>
> I was alone as I started my backbends. I did one, came down, then came up for two more. Then I did three more, trying to rock to standing, resting in between.

Figure 6.5
A cybershala resource for learning kapotasana (lrockwood 2006). (*Screenshot by the author*)

For the past few days, I've given up on getting any real lift and I haven't been thinking much about straight legs either. To be perfectly honest, I've been pretty relaxed about the whole thing, just doing the rocks and not getting too attached to the outcome. [...]

The "magic moment" happened in my second attempt at rocking. I was doing my rocks and just thinking about lifting my chest. Nothing heroic, just wondering if I could lift it maybe three inches higher.

On the second rock, something shifted and I felt my legs "connect" and my hands came off the floor for an instant, then I ran out of inhale and came back down. I need bigger lungs!

I've managed to get my hands to lift off the floor before, but it was through pure momentum (rocking). This felt different. It felt like my legs took control of the movement, like a baton being passed in a relay. I recognise the sensation—I've felt a hint of it when DR has helped me stand up but I never felt like I "owned" it.

I'm making no predictions. It could have been a fluke. Since the "chest lift" visualization seems to be working for me, I think I'll keep with it. (Kaivalya 2010)

Yoga bloggers are engaging in what Jaida Kim Samudra calls "thick participation," a riff on Clifford Geertz's "thick description" (i.e., richly detailed ethnographic writing that interprets social discourse [Geertz 1973]). Thick participation is a fieldwork method that centers on bodily practice rather than on talking or observation; it generates "cultural knowledge recorded first in the anthropologist's body and only later externalized as visual or textual data for purposes of analysis" (Samudra 2008:667; cf. Willis 2000). In her research on White Crane Silat, a Chinese Indonesian martial art, Samudra faced the challenge of writing about the cultural meaning of kinesthetic experience in a community where practitioners rarely talk about what they're doing. Teachers discouraged efforts to verbalize about the learning process, telling Samudra that "it would interfere with the direct transmission of silat" from body to body (669). In ashtanga yoga, as in White Crane Silat, cultural knowledge is "deeply embodied and often not transmuted into semiotic code" (666). However, as Samudra demonstrates, that doesn't mean that knowledge is permanently "trapped inside the bounded, individual body, always unknown to others" (677); members of kinesthetic cultures share common knowledge and communicate through their bodies.

Yoga bloggers are taking on the same task that Samudra addressed in her scholarly writing: how can one communicate one's kinesthetic experience *without* bodily presence and a shared sensorium? They have adopted exactly the same tactics that Samudra advocates for scholars writing about kinesthetic cultures: first, attempting to "linguistically record the minute details of one's bodily training" even when this is explicitly discouraged in traditional transmission (2008:670); second, being attentive to one's own internal bodily sensations in order to better comprehend other practitioners' experiences (674); and finally, creating "somatic narratives," which comprise both "the series of actions narrated by bodies during limited frames such as practice sessions, performances, or competitions" and "the stories people tell about what happened to and with their bodies during specific events" (674). The Reluctant Ashtangi's backbend post employs all three tactics: Her writing is minutely detailed, she experiments with metaphors to express internal sensations, and her post takes the form of a somatic narrative.

As I read more yoga blog posts and watched more videos, my body shifted in my computer chair as though operated by remote control: back straightening, shoulder blades sliding together, legs subtly rotating in hip sockets, toes spreading to grip the floor for a vicarious backbend. (See grimmly2007 2010a, ▶ 6.4.) I heard myself breathing. Just as I had learned to reenact my teacher's physical adjustments while alone on my mat—the hand drawing my hip back, the foot nudging the angle of my own foot on the floor, the arm stopping me from taking my legs past vertical in a headstand—I experienced the blogs and videos through my accumulated "sensational knowledge" (Hahn 2007). As Samudra puts it, "The communications of the body can be verified even when not encoded into language because they work in practice" (2008:667). It was very much like the experience of listening to music that I knew how to play. When I got up and tried a few backbends, I was almost surprised that I still couldn't do a drop-back to the floor—though I did feel inches closer than I'd been a few hours earlier at the yoga studio.

I also had the uncomfortable feeling that I might be cheating on my teacher. In the United States the ashtanga teacher–student relationship generally does not come freighted with the full weight of the Indian guru–disciple tradition (*parampara*), but teachers explicitly trace their own lineage back to Sri K. Pattabhi Jois. Many practitioners make annual trips to Mysore to study at Jois's Ashtanga Yoga Institute; this is how teachers eventually earn "Authorized" or "Certified" status from the Institute (SKPJAYI 2009). Ashtanga students build up their practice asana by asana, with the teacher determining when a new asana should be added. Students are not supposed to start experimenting with advanced asana of their own accord any more than classical piano students are supposed to work on virtuosic Liszt etudes when their teacher has assigned them Mozart sonatas—with the important difference that the yoga students may be risking physical injury. Ashtanga practitioners generally treat their teachers with respect and deference, grounded both in *parampara* and in the cumulative trust developed through a physically intimate transmission process. For instance, your teacher might stand in front of you, pelvis to pelvis, and hold you by the waist while you raise your arms overhead and lean backward until your hands land on the floor. Less terror-inducing but equally important, teachers know what constitutes productive physical discomfort and what might be an incipient hamstring tear.

On the other hand, the structured nature of ashtanga makes it particularly well suited to independent practice, amateur-to-amateur pedagogy, and online discourse among a dispersed community of practitioners. Students are encouraged to practice six days a week; if costs or time restrictions preclude six weekly visits to a shala, then they are supposed to practice at home, much like musicians practicing between lessons. Ashtanga is also eminently portable. Anyone who knows primary series can walk into a Mysore-style class anywhere in the world and join the local community of practice. Everyone who practices second series knows exactly what it means when I say that I am working on kapotasana; they have their own sensational knowledge of that asana. If the asana came more easily to them than to me, then they also have a visceral understanding of the differences between our bodies. As any musician or dancer will tell you, sharing a specific repertoire of embodied practice lays the groundwork for powerful, intimate social connections: "Technique binds strangers to one another, transforming contingent arrangements, over time, into relational infrastructure" (Hamera 2007:59).

Browsing YouTube videos of ashtanga backbends quickly led me to "grimmly2007," who had uploaded about 300 videos so that he could embed them in his yoga blog. Grimmly is an ashtanga yoga student without a teacher—an impossible contradiction to many practitioners, but one that is apparently getting more possible all the time.

Anonymous said . . .
    Whoah! You've woken up the Ashtangisphere Grimmly! All this dialogue;)

Grimmly said . . .
    Ahhh, the ashtangisphere never sleeps, one eye is always open:)
    But yes, all these Home Ashtangi's out there commenting(
    Thank you everyone). We've been quiet too long, intimidated by the Shala
    and Mysore ashtangi's.
    Home Ashtangi's of the world unite!
    Let your voice be heard,
    we practice too!
    say it with me,
    "WE PRACTICE TOO" lol
    (comments on Grimmly 2010: part 7)

An overwhelming number of yoga blogs, videos, Facebook updates, Twitter feeds, and other forms of online social media now constitute a "cybershala" of ashtanga yoga practitioners—many who work with teachers regularly, others who are cultivating a practice as "home ashtangis" (cf. Finnegan 1989 on "hidden musicians"). Grimmly's blog offers a rich portrait of one person's "home ashtangi" experience. For the rest of this section, I will present my discussion of the cybershala in dialogue with his posts.[7]

> I'm by nature quite solitary and have never found practicing alone an issue. It was quite a surprise, then, to wake up one morning recently and realise that I was part of an Ashtanga community, the Cybershala. [. . .] Looking at my blog counter today, I see that I've received over thirty thousand visitors and am averaging a hundred and fifty hits a day. From the feedjit application I see that visitors come here, stay a few minutes and then head off to one of the other blogs mentioned on my page, sometimes coming back a few minutes later before flitting off somewhere else. I have this image of bees buzzing from flower to flower. (Grimmly 2010: part 17)

Grimmly lives in the United Kingdom and works as a repairer of woodwind instruments; he also has some past experience teaching English and philosophy. In early 2007 his flat was burgled and seven saxophones were stolen. This incident made him so angry, and then so irritated with his own anger, that he decided to take up some form of meditation. In the course of reading about meditation practices, he learned that "a lot of meditators were also doing yoga," so he looked for a book at the library and found Tara Fraser's *Total Astanga* (Fraser 2006). As an overweight 43-year-old man, he was a bit embarrassed even bringing the book up to the circulation desk. "Going to a yoga class wasn't something I even considered. A guy here, outside London, might think about going to a gym to get in shape but not a Yoga class, probably not even an aerobic class" (Grimmly 2010: part 1). Moreover, since he was trying out yoga as part of a meditation practice rather than as an exercise regimen, it seemed reasonable to do it alone at home.

Grimmly began learning the sequence of asana from the book, practicing every morning before work, and "of course at the same time I started to Google. [. . .] I remember seeing that Youtube video of the guy practicing in the temple and being blown away by it (still am) http://www.youtube.com/watch?v=Hu9Sq1RvuoA, was this what I was doing?"[8] Soon Grimmly bought an instructional DVD: "It was a good job I bought the video when I did, [the teacher] would give lots of advice for practicing safely, if I'd just carried on with the book I would probably have injured myself" (Grimmly 2010: part 2). Next he picked up David Swenson's practice manual, at which point "I began to think of my practice as an Ashtanga practice. I began to take it all much more seriously. You open the cover and there's K pattabhi Jois smiling at you. On the next page there are some early pictures from 1975 of the guys who first practiced Ashtanga in the West and a little box called A Living Tradition" (Grimmly 2010: part 3).

Grimmly started his yoga blog in the summer of 2008, after about a year and a half of practicing at home alone six days a week. The blog was intended to help him document his progress and to compile web resources on the "jump-back," a transition between many asana: Essentially, it consists of moving from a seated, cross-legged position to a low pushup position by pressing up on one's hands, curling into a ball, and shooting one's feet to the back of the mat. (Grimmly's 57 blog posts on the jump-back hint at how much is missing from this description; see grimmly2007 2010b, ▶ 6.5.) Although Grimmly had still never attended a yoga class, his first post demonstrates how much ashtanga discourse he had already absorbed from books, DVDs, and the cybershala:

> I've heard it said, and been told that the Jump back is not so important and not something to get hung up on, that it will come with time etc etc. And I agree with this. I know and accept that it's just one aspect of the practice and there are many ways/variations of the jump back that allow you to string the asana together. And I'm aware that Nancy Gilgoff [a famous teacher], supposedly, never employed the jump back.
>
> BUT
>
> I like the jump back!
>
> I want to jump back!
>
> I WILL jump back!
>
> I like how it links the asana. I like how the asana flow together when you can jump back out of one asana and jump through into the next. Ashtanga is a flowing practice, so what's wrong with focusing on the flowing aspect.
>
> We shouldn't have to defend our keenness to master it.
>
> (Grimmly 2008a)

This first post strikes a tone that remains consistent over the next three years and more than 500 posts. Grimmly often cites received ideas about ashtanga, thereby both placing himself within a community of practitioners and preempting corrective advice by establishing his own credentials. He also invokes a "we" consisting of fellow students—not only other hidden "home ashtangis," but those who are working with teachers who admonish them not to obsess about the jump-back. Most importantly, he asserts that his practice is his own, pushing back against traditional authority. While Grimmly's blog tracks his increasing awareness of and respect for the ashtanga lineage, he "wasn't in the market for a new philosophy or belief system. The practice was enough, still is" (Grimmly 2010: part 4).

Grimmly engaged with the thorny issue of authority in a post discussing various assertions as to what ashtanga yoga is "about" (e.g., asana, breathing, "the realization of one's true nature"). He was inspired partly by hearing from a friend "who told me that an Authorised teacher had written to her saying that practicing in the evening was not an honest practice" (comment posted on Owl 2009). Once again, Grimmly asserted each student's earned and inalienable right to his or her own practice:

But you know what, this practice, it's about pretty much whatever I want it to be about. Ashtanga, it's out there, authorial authority died a death a long time ago.

Who knows what the Korunta was *about* before it was taught to Krishnamacharya (assuming it was) or what it became about when he taught it to Jois. . . . And is that what the practice is about to Sharath [Jois's grandson] or is it different in his mind, we shall see in time. It certainly became about something else to the Western students who brought it back with them from Mysore with their Western paradigms and preconceptions. Notice the focus on anatomy, or the post Freudian/Jungian/Kleinian/Lacanian discourse that gets woven into talk about the practice. The influence of New Age philosophies, NLP, cognitive theory. . . .

In the end a practice is practiced. It's practiced by somebody and this is a hard, intense, demanding practice. Anyone who practices it six days a week, year in year out, earns the right to determine for themselves what the hell this practice is about. (Grimmly 2010: part 16)

Here Grimmly draws attention to a paradox of ashtanga transmission: While teachers are indeed accorded "authorial authority," traditional pedagogy focuses on practice rather than preaching. Teachers often cite Sri K. Pattabhi Jois's statement that ashtanga is 99 percent practice and 1 percent theory (a precept that seems a bit at odds with the reams of verbiage in the ashtanga cybershala). In Mysore-style classes, the room is quiet, except for the students' breathing and the teacher's occasional verbal explanations addressed to individuals. No one is declaiming to the class about "mindfulness" in the manner of stereotypical New Age yoga teachers. However, all authorized teachers and many experienced students are also committed to particular metaphysical and ethical principles: the "eight limbs" of yoga codified in the Yoga Sutras (usually attributed to Patanjali, circa second century BCE). These practitioners are endlessly frustrated by those who treat yoga as just another form of physical exercise or as a facile path to "spirituality." But when they express their convictions, they are often hindered by their reliance on terms that skeptical newcomers find indistinguishable from generic "spiritual" language.

In the cybershala, practitioners negotiate these issues by translating each other's experience into or out of marked language (cf. Hamera 2007:175–179). For instance, when Grimmly wrote that he had originally decided to "give some yoga a go" because he was angry that his flat had been burgled, a commenter declared that he had apparently experienced "sufficient new suffering to create intense desire to quiet mind." Grimmly's reply pushes back against this translation into spiritual-ese:

I wouldn't call it 'intense suffering' but rather a growing annoyance with still being peeved about whole thing and read 'give some yoga a go then' for 'intense desire' at wanting to get back to my general easy going, laid back disposition . . . rather than 'quiet mind'.

But I guess 'suffering' 'intense desire' and 'quiet mind' sell more ink.

(comments posted on Grimmly 2010: part 2)

It's hard to imagine hearing this kind of exchange during changing-room conversation at a yoga studio, but it is not at all unusual in the cybershala. Nor did the commenter seem to take offense; she or he and Grimmly continued to read and comment on many of each other's blog posts. The distancing effect of virtual discourse is often remarked on in discussions of online incivility; however, it is equally possible for web-mediated distance combined with shared visceral experience to create the conditions for dialogue that would otherwise never take place (cf. Miller 2008b, Ch. 6).

As Grimmly developed his home practice, he gradually began to "talk the talk" of ashtanga yoga—for instance, referring to the *bandhas* (often glossed as "energy locks") instead of using the language of Western physiology and gym culture ("abs," "pelvic floor muscles," "core muscles").[9] However, some of the choices he made as a student-without-a-teacher continued to challenge ashtanga orthodoxy—or, rather, to draw attention to its practical flexibility. For instance, when Grimmly blogged about his decision to begin learning the second series of asana (the "intermediate" series), one commenter told him that he should not be learning any intermediate asana before he could stand up from a backbend: "Then and only then you start to add intermediate to your existing primary. Your teacher would give you each new asana as he saw your progress. [. . .] Traditionally in India, yoga has been learned from teacher to student, not from a book or video. It's really not right to decide to give yourself postures" (Sophia, comments posted on Grimmly 2008c).

When making such claims about traditional practice, ashtanga practitioners often invoke the ultimate authority: "how it's done in Mysore," at Sri K. Pattabhi Jois's shala. However, in this case another commenter offered evidence that the teaching method in Mysore has changed over time. Ursula, a woman from Germany, wrote, "When fewer people came to Jois, the second series was teached after one months of practicing the first series. There were not these restrictions like now. [. . .] And I still want to add something. I was in Mysore this year. I am not able to drop down into urdhva dhanurasana [the backbend under discussion], nor am I able to come up from this pose. I was given pashasana [the first intermediate pose] there." When another commenter suggested that this experience might have been "an aberration," Ursula responded by invoking the higher authority of her own bodily experience: "Sorry to write this, but what I see is that people hanker for rules, because there is so much insecurity. People do not trust their bodies anymore. [. . .] The poses of the second series helped me to improve UD [urdhva dhanurasana].

Why should I not do these softer back bending which are good for the back? Only because there are rules, nobody really knows who invented them. [. . .] I listen to my body." In a response, Grimmly suggested that some founding texts could be considered authoritative, but emphasized the variability of authorized teachers' interpretations of these texts. He ultimately reinforced Ursula's argument: "It's my body and it's my practice. Mine, not yours, it's no longer even Guruji's [Jois's] or Sharath [Jois's grandson]. Like a book that once written is up to the reader to experience and interpret for him- or herself, not the author or the university lecturer or PhD student. Show me two practices the same" (comments posted on Grimmly 2008c). This discussion continued at length, showing how the print medium and time-delayed norms of comment threads have encouraged the development of cybershala discourse. Discussants can take time to craft their replies and cite their sources, which in this case revealed the fluidity of "official" ashtanga pedagogy in Mysore.

After a year and a half of home practice, Grimmly finally decided to try attending a Mysore-style class at a shala. He went two Sundays in a row and was "blown away" by the physical adjustments he received from the teachers there:

> L. just kept turning me further and further into it until I could grasp my wrist rather than just my fingertips. Felt like if she let go my legs would go spinning round and round in cartoon fashion. [. . .] Seems so generous. It's one thing to stand at the front of a room and say do this, do that but to get down on a sweaty mat and help our sweaty bodies into these asanas just seems such a generous selfless act. (Grimmly 2010: part 15b)

But a week later, in a post titled "Owning my practice," he explained why he doubted he'd return:

> I've been working on all these elements for the last couple of weeks. . . . Thing is, my practice doesn't feel mine anymore. Or less mine. [. . .] All the time it's just been me on my mat, alone in a room early each morning, my practice. It's followed my mood and inclination, will, desire, frustration, stubborn determination, whatever. Somehow now, after visiting the Shala, it feels a little like I'm practicing for someone else, my teacher? I need to work on this or that, improve this or that. Those elements to work on didn't come from me, didn't arise in me. [. . .] Don't get me wrong I'm so very grateful for the attention, the adjustments, advice, suggestions It's just that each morning this last week it's felt a bit of a chore, my hearts not been in it. I feel more distant from my practice, less involved. (Grimmly 2008b)

It's clear from other posts that Grimmly has developed his practice using books, famous teachers' DVDs, YouTube videos, other students' blogs, and any other media resources he can find. He often writes about insights gleaned from these sources. Nevertheless, the "live" teaching at the shala somehow alienated him from his practice. Grimmly's experience contradicts a core assumption voiced by many of the online teachers cited in this book: that the lack of individual, real-time, physical feedback in web-based pedagogy represents a huge shortcoming. While Grimmly benefited from the physical adjustments he received at the shala, he was willing to forego them in order to maintain a sense of agency and responsibility for his own development: practicing for himself instead of a teacher.

Yet while Grimmly never affiliated himself with a shala, he certainly does not practice yoga in isolation. Grimmly practices in the cybershala, and he participates in an international network of amateur-to-amateur pedagogy (see Figure 6.6). His blog is an integral part of his practice, not simply a form of documentation:

> The Blog quickly became addictive and started to take over my practice. Once I'd caught up with the things I had already planned on posting I would find myself in the middle of a Jump back thinking 'Oh this is interesting must post this' or 'That was better, must video it for the blog'. This was both good and bad, distracting but motivating. [. . .] I wonder, if I hadn't started this blog, if I'd still

Figure 6.6
A Feedjit widget map showing the locations of recent visitors to Grimmly's blog as of August 24, 2010 (Grimmly 2010). (*Screenshot by the author*)

be practicing now or if I would have become frustrated and given up. If you're practicing alone, I highly recommend it. (Grimmly 2010: part 10)

It's not just that I was able to devote myself to the practice, but also that I had the time to work on this Blog and use it to reflect on my practice. I was able to take Jump back clips from all my DVDs, break them down into screenshots and turn them into slideshows. [. . .] I'd break down my own Jump back as well as Kino's say, and run the two slide shows side by side and try to work out what she or Lino or Sharath were doing right that I wasn't. (Grimmly 2010: part 13)

Grimmly and his fellow cybershala practitioners are creating new transmission modalities for ashtanga yoga, from reflective writing grounded in "thick participation" to side-by-side slideshows that might reveal hidden traces of corporeal knowledge. Meanwhile, they are working through their individual relationships to traditional authority and virtual community: What does it mean to "own" a traditional practice? Is it ever possible to practice "alone"?

As Grimmly writes, "A practice tends to be social, it has a history, it has a world. Strange how the personal and the social come together here on the mat. The practice then is never private" (Grimmly 2010: part 7). In 2011, I happened to send him a draft of this chapter just after he had written a "Gone Fishing" post that stated that he would be taking a summer hiatus from blogging. In response to my email and to the comments on his farewell post, he returned to regular blogging, with a post titled "Back so soon?"

Well this is embarrassing, a little awkward, seems I'm not going to be taking a long summer break after all. [. . .] I blog, however poorly, it's helped me with my own practice and seems it can be encouraging/motivating to others at times, even inspiring, just as others blogs have been inspiring to me. [. . .] I've also noticed how absurd it was, there was me troubled by the insular nature of practice and in response I go and make it more so by quitting the social aspect of sharing and supporting, stupid yogi. (Grimmly 2011)

Ashtanga yoga offers an intriguing test case for A2A pedagogy because this practice seems fundamentally incompatible with web-based transmission. The web privileges written text, sound, and sight. Ashtanga yoga teachers minimize verbal instruction and explanation, and once students grasp the basic form of an asana they aren't supposed to learn visually, either. (Each asana has an associated gazing point, which precludes learning by looking at other students during group practice. Ashtanga teachers also avoid teaching in rooms with mirrors.) However, both economic pressures and the desire to promote the practice have led many teachers to produce and market books and videos, implicitly encouraging visual and verbal learning. Of course, these books and videos were produced by credentialed experts, "certified" or "authorized" ashtanga teachers; otherwise, who would buy them? But in the Web 2.0 era, it became possible for a "home ashtangi" like Grimmly to reach thousands of practitioners spread around the world—and to decide that he preferred practicing in the cybershala to joining a brick-and-mortar shala with a flesh-and-blood teacher.

This gradual paradigm shift in ashtanga yoga transmission might shed light on similar changes in the transmission of performing arts traditions that rely on a lineage of teachers and students, body-to-body pedagogy, and a codified repertoire or fundamental skill set. As Nasir Syed has observed with respect to online Hindustani music lessons, "The master–apprentice relationship . . . is being renegotiated" (Cooley, Meizel et al. 2008:104). The ashtanga cybershala draws attention to competing forms of authority invoked by students of such traditions: authoritative institutions or teachers versus "listening to my body" or "owning my practice." Who has the right to represent a traditional practice, particularly when many respected practitioners cannot claim authority based on ethnicity or cultural heritage? When does repetition with a difference become repetition with too much difference? These are familiar topics of discussion among scholars of traditional arts around the world; now they are also common topics in A2A discourse.[10]

CONCLUSION: SOME KEY TRAITS OF AMATEUR-TO-AMATEUR PEDAGOGY

These brief excursions into three A2A learning contexts demonstrate the diversity of approaches in contemporary A2A pedagogy. To summarize:

Nate Torres's nine conga lessons are organized into a structured curriculum and focus on fundamentals: the basic techniques and foundational rhythms that he hopes will get aspiring congueros off to the right start. Like David Taub and Nate Brown (see Chapter Five), he has adapted the style and structure of traditional private lessons to suit the YouTube format. When you study conga with Nate Torres, you meet him in his living room (conveniently located on You-Tube), sit face to face, and work your way through a graded series of lessons. He exhorts you to practice, practice, practice and to be patient with yourself.

Pianojohn113 found traditional private piano lessons alienating. He therefore creates lessons that encourage students to immediately start playing songs that they love. When you study piano with pianojohn113, he serves as a medium, offering you access to the creative process and embodied performance experience of a beloved musician. You never see pianojohn113's face—just his hands on the keyboard. He talks you through the music riff by riff, giving you time to let your hands feel out how to follow and inhabit the motions of the hands you see on screen. Magically, music that you know and treasure comes alive under your fingers. In some respects, learning from pianojohn113 is a lot like playing *Guitar Hero* or *Rock Band*, but without the programming shortcuts that allow beginners to produce perfect audio output.

Grimmly did not set out to teach ashtanga yoga at all; he started his blog intending to document his progress with one particularly challenging physical move, the "jump-back." The blog format offered him an easy way to compile online resources and to reflect on his learning experience. Before long, though, he began to receive questions and comments from other yoga practitioners. As he responded to them, he became an active member of the cybershala and a source of inspiration and guidance for other "home ashtangis." While most cybershala participants do seem to work with a "live" teacher regularly, the comments on Grimmly's blog show that he has inspired many people to "own their practice" and develop it through A2A resources.

Despite the differences among these three teachers, their chosen traditions, and their methodologies, these case studies highlight some core traits of amateur-to-amateur learning. First, *A2A destabilizes teacher–student relationships.* Roles shift quickly: Students are pressed into duty as teachers, and teachers are sometimes treated as students. Established ideas about expertise, authenticity, and authority become subject to public debate grounded in idiosyncratic individual experience. Online social media platforms also encourage people to blur the lines between their social roles; the personal and the pedagogical become intertwined.

This is not to say that A2A participants simply reject structure, hierarchy, and hegemony. The examples in this chapter show that A2A pedagogy often results in implicit or explicit challenges to institutional authority, but they also demonstrate a second key trait: *A2A discourse flourishes in connection with practices that involve canons, codified techniques, acknowledged master practitioners, and instruments that shape practitioners' embodied experience.* These established structures give rise to shared knowledge and visceral experiences that in turn build a scaffold for meaningful online discourse.

Third, *A2A pedagogy inspires creative experimentation with available media technology and expressive styles.* A2A participants constantly face the challenge of communicating corporeal knowledge using words, pictures, and video. Pianojohn113 adds screen annotations to his videos, reinforcing his verbal explanation and physical demonstration. Grimmly posts videos that are radically slowed down (to reveal the mysteries of the jump-back) or sped up (to offer a quick reference for learning a long sequence of asana). Then he counters this emphasis on visual learning by posting videos of himself and others practicing blindfolded (Grimmly 2009). Many A2A participants also experiment with "somatic narratives" (Samudra 2008) and evocative metaphors that are intended to help readers understand, identify with, or reenact particular kinesthetic experiences. Their use of language is "bound up with practices in the world, where the point of verbalizing is . . . to produce sensuous reconstructions in the reader's mind" (Willis 2000:22).

Fourth, *A2A pedagogy generates a sense of mutual obligation, emotional investment, and social connection among participants.* As Nate Torres wrote in response to thanks from an online conga student, "I really enjoy all of this and I'm always here to help and aid those who are seeking it . . . I almost feel like I'm gaining a little 'family' here" (comment posted on prpapito3000 2007c). The same could be said of many learning communities, but it seems notable that these social and emotional connections play such an important role in a situation where students could just as easily be anonymous, casual, and transient. The fact that no money is changing hands seems to enhance this phenomenon: A2A students greatly appreciate their teachers' willingness to share their expertise for free, especially in a context where "performers with commercial ambitions—and, at times, corporate sponsorship—will use the cachet of the 'homegrown' and the 'grassroots' . . . to advance their budding careers" (Salvato 2009:69). Devoted amateurs like pianojohn113 win loyal students, who then defend their teachers against the incivility and "trolling" that often plague public comment threads. A2A teachers also make liberal use of the comment/reply functions built into their chosen media platforms, publicly demonstrating their commitment to interaction with individual viewers/students/fellow practitioners. These exchanges can continue long after an A2A teacher has stopped posting new content online. As I write this chapter it has been over two years since Nate Torres uploaded the last of his nine conga lessons to YouTube; nevertheless, he continues to respond to viewers' comments and questions.

Finally, because A2A participants are dispersed and diverse, *A2A discourse constantly invokes competing standards, definitions, norms, and authorities.* This diversity is constrained by factors such as language, economic resources, technological infrastructure, and government censorship. Nevertheless, the examples in this chapter reveal considerable diversity among participants—for instance, in terms of age, gender, geographic location, profession, educational background, religion, ethnicity, national origin, and previous experience with the practice in question. Should you cup your hand to produce a slap tone on the conga? Is it more authentic to play exactly what Billy Joel played on a recording, or to emulate his creativity and never play a fill the same way twice? When is an ashtanga yoga student ready to start learning the intermediate series of asana, and who is qualified to make that assessment? Is it fair for a copyright holder to order people to stop teaching each other popular songs? These kinds of conversations might never take place if A2A participants were learning from masters on the holodeck rather than learning from each other across a network of living rooms.

Endgame

If we live a "virtual condition," . . . it is not because we are condemned to the fake, but because we have learned to live, work and play with the fluid, the open, the potential. (Ryan 1999:94)

From *Grand Theft Auto* to the cybershala: a seemingly long and improbable trip. What could the GTA player and the yoga practitioner possibly have in common, apart from the cultural and economic capital required to use interactive digital media in their leisure time? The short, personal answer is "I do both"—as many *Guitar Hero* players have answered when someone asks, "Why don't you pick up a real guitar?" In the winter of 2005–2006, when I started this research project, I had an afternoon routine. I'd spend an hour playing *GTA: San Andreas*—making several attempts at a tricky mission, trying out various radio stations in an effort to improve my focus, shaking out my sore wrists during pauses in the action, and writing some fieldnotes at the end of the session. Then I'd bundle up for the below-zero Edmonton winter and walk to my yoga studio, where I'd join a room full of people who were all working through a set series of postures—all of us as choreographed as my GTA avatar, but with a lot more variety in our technical execution of the repertoire. At home, in front of my basement TV, I struggled with the interface: I'd hit the wrong button and CJ would do something I hadn't intended, like stealing a car instead of getting on his bike. In the studio, my body was equally resistant to my intentions, and sometimes equally surprising: the accidental carjack mirrored in the accidental headstand.

I've always been deeply involved with communities of practice and the transmission of tradition, from childhood ballet and piano lessons to my dissertation project on participatory singing from the *Sacred Harp* tunebook. The experience of practicing scales, repeating *tendu* exercises at the ballet barre, singing the same songs a thousand times, and sharing visceral common knowledge with other practitioners has undoubtedly influenced my approach to interactive digital media. I've never seen video-game players as automatons with glazed eyes glued to a screen; instead, I recognize the same focused attention, intense engagement, and deep pleasure I've experienced when playing from a musical score or performing challenging choreography. Game rules, theatrical scripts, musical scores, choreography, athletics, and traditional repertoires of all kinds offer the structural constraints, practical challenges, and variable outcomes that engender "flow," embodied knowledge, and a sense of common experience with others. With or without the involvement of digital-media technology, all of these forms of "playing along" at times rely on virtual performance: exploring the boundaries of one's potential, or "approaching the actual without arriving there" (Boellstorff 2008:19).

So what makes playing along with interactive digital media distinctive? In a word: access. Playing along is a privilege, and achieving its rewards has historically required time, money, and the development of face-to-face relationships. Consider the resources required to play piano four-hands arrangements, perform a classical dance repertoire, act in a theatrical production, join a rock band, or participate in team sports. Such pursuits generally entail in-person, real-time contact with other people: teachers, fellow participants, audiences. Would-be participants have to coordinate their schedules, pay for lessons and special equipment, and submit to the traditional authority of teachers or coaches. Then they have to go through the long, repetitive, often physically painful process of training their bodies to move in new ways, practicing a repertoire until it becomes second nature. This assumes that they can actually gain access to the community of practice in the first place—even setting aside matters of time, money, and geographical location, many communities present explicit or implicit barriers based on gender, age, class, race, ethnicity, sexuality, religion, language, body shape, or physical ability.

Interactive digital media certainly can't be credited with dissolving all these barriers to access. It still takes a certain amount of time, money, physical ability, and commitment-to-practice to become a competent *Grand Theft Auto* or *Rock Band* player, and identity factors still shape participation in web-based communities. The research presented here reflects that fact, most notably with respect to gender. All of the leading roles in this book are played by men (with the important exception of my own role). While female practitioners make appearances in every chapter, they were underrepresented in most of my virtual and actual research contexts. When I recruited players to complete online surveys or to volunteer for gameplay/interview sessions, the vast majority of respondents were male—even though I did some targeted recruiting of female players through my own social networks. Men vastly outnumber women on most gamer messageboards. Online or offline, men are more likely to identify themselves as "gamers" and to feel that they have something to say about the subject, despite the fact that American video-game players are now 40 percent female

(ESA 2010). Rock instrumentalists are also predominantly male, a fact not lost on the guitar and drum teachers I discussed in Chapter Five. Both David Taub and Nate Brown have taken steps to address this imbalance, by recruiting female teachers and online forum moderators and responding swiftly to students' concerns about sexist content in forum posts or videos. Still, for now, men substantially outnumber women at both OnlineDrummer.com and Next Level Guitar.

Nevertheless, it's clear that both more people and more *kinds* of people can play along with the games, performance genres, and repertoires circulated via interactive digital media—not everyone, not everywhere, but millions of people around the world. A gamer messageboard like ScoreHero.com might present a gender split nearly as lopsided as that of the *Guitar Hero* tournament I described in Chapter Four, but access to forum participation is relatively gender-neutral; I certainly felt more comfortable posting to online forums than visiting the traditionally male spaces of gamer tournaments or guitar stores. Moreover, I could find online communities, teachers, and repositories of practical information via web searches, rather than breaking into an exclusive social network or having to work around people's assumptions about what I "should be" seeking (e.g., a "girl game" or a "grrrl gamer" forum, a female guitar teacher, acoustic guitar lessons rather than electric, voice lessons instead of drum lessons). By the same token, the yoga "cybershala" proved accessible to Grimmly, an overweight 43-year-old man trying to develop a yoga practice, although he lived in an area where even checking a yoga book out of the library felt like an embarrassing gender transgression: "Going to a yoga class wasn't something I even considered" (Grimmly 2010: part 1). While Grimmly was uploading backbend videos, YouTube was also offering Freddie Wong a platform for performing rock drag to an audience of millions; online, he could parody metal masculinity and bait homophobic and racist commenters without worrying about a visceral back-alley confrontation after the show. And no matter who you are, playing *GTA: San Andreas* or *Rock Band* is far more accessible than starring in a hundred-hour theatrical production about Los Angeles gang life or performing a huge repertoire of songs with a touring band.

When real-time and in-person interaction are optional rather than essential, communities of practice can extend across time, space, and some cultural and technological boundaries. Social media platforms that accommodate time-delayed interaction support participation across time zones; they also make room for people with limited Internet access or slow connections, and for those who require extra time to understand or communicate in the language being used. People who routinely use these media have recognized and built on this potential for broadening one's audience, student roster, customer base, or virtual community. Sometimes they do so by design, as when commercial game developers or professional teachers strive to create games or guitar exercises that "anyone can play"—a mantra David Taub repeats in countless lessons. Sometimes virtual affinity groups seem to grow of their own accord: Nate Torres, Pianojohn113, and Grimmly made no special effort to develop an audience, and they were all astonished when they began receiving comments from students or fellow practitioners all over the world. People found them through serendipitous keyword searches, mediated by the institutional patronage and proprietary search algorithms of companies like YouTube, Blogger, and Google (which now owns both the others). Clicks begat clicks, views begat views, and more people began practicing conga, piano, and yoga in distant corners of the globe.

Shifting parameters of access inevitably change transmission processes, which in turn change the nature of what's being transmitted. Sometimes these changes invoke earlier practices and values, as when *Guitar Hero* reinvents amateur parlor music as schizophonic performance: Why just listen to a song when you could play it instead? Digitally mediated transmission can also rock the foundations of traditional authority, as when a yoga practitioner learns from mass-produced media and fellow students in the cybershala rather than developing an intimate master–apprentice relationship with a particular teacher. Knowledge of countless subjects and practices is increasingly dispersed and distributed; concentrated authority is becoming diluted and hierarchical relationships are flattening out (Gee 2004:85–87).

While my amateur-to-amateur case studies show how quickly an unknown amateur can become a public teacher, such transitions occur by choice, not necessity. Interactive digital media offer unprecedented options when it comes to navigating public and private spheres. It has become a truism to observe that the Internet age is eroding privacy, but it is equally important to recognize that the same technology has made it possible to play along in private without playing alone. This, too, expands access, by eliminating the risk of public embarrassment. People feel free to play against type, and sometimes to fail spectacularly on the first few tries. *Grand Theft Auto* players and cybershala yogis alike can explore the wealth of online resources related to their chosen practices and experience a sense of visceral common knowledge shared with uncountable unknown others—all without ever taking their own practice public. "Hidden musicians" have new opportunities to make themselves known—by posting their living room performances to YouTube or creating an A2A lesson series—but they can also engage meaningfully with others' contributions without becoming public performers. One of the volunteer moderators for the Next Level Guitar forum owns several guitars, practices at home every day, vets thousands of messageboard posts, but never plays in public. He told me, "I have such a terrible problem with stage fright—I just never got over that, so I just kind of play for myself and in my office here in my own little world. My wife and my kids will listen. I don't specifically go out of my way to perform for them."[1]

Interactive digital media also limit or close down certain kinds of access—most obviously, access to a shared sensorium and interpersonal physical contact. Sherry Turkle has addressed the implications of these limitations in her recent book *Alone Together*. She argues that interactive technology and social networking platforms undermine genuine human intimacy, leading us to "expect more from technology and less from each other" (2011). But as this book has shown, these same limitations have also spurred the development of media products, platforms, and practices that create visceral, intimate connections by making the most of images, sound, and typed characters. As these technologies have become ubiquitous, people have pressed them into service even when they seem like a terrible match for the task at hand. Thus gamers, musicians, dancers, and athletes engage in virtual performance and creative trans-sensory interpretation, finding ways to communicate actual embodied experience in a virtual medium. Technology may catch up; the holodeck may finally arrive. But meanwhile, people are discovering what they can do without it. Playing along allows people to experience intimate connections with other practitioners even if they never meet face to face.

Developing medium- and platform-specific communication strategies, becoming invested in particular genres of virtual performance, and reaching out to the broadest possible online public also means dealing with commercial and legal constraints. As David Taub learned, guitar teachers can teach all the copyrighted songs they want in private lessons, but if they want to build a learning community on YouTube, they had better stick to technical lessons or risk account deletion. Unless *Guitar Hero* and *Rock Band* players have specialized hacking and modding skills (and expertise in musical transcription), they can only play songs that have been licensed, coded, and distributed by Harmonix and Activision. Similarly, only a tiny percentage of *Grand Theft Auto* players have the programming skills required to create and play through new storylines with their avatars. While interactive digital-media companies sometimes release sophisticated authoring tools to support the creation of user-generated content—as Harmonix did with the 2010 "Rock Band Network" initiative—they also generally maintain some control over that content: Play with my tools, play by my rules. These limitations are significant, and they present an important counterweight to celebratory narratives about ever-expanding access, the democratization of creative production, the end of authorship, and the egalitarian nature of peer-to-peer interaction.

But while it would be a mistake to ignore the role of commercial forces in structuring the practices described in this book, it would be equally myopic to dismiss ordinary players' practices as nothing but compliant consumption. As I've discussed my technoculture research with colleagues, students, friends, and the occasional journalist over the years, I've discovered that many people assume I must be most interested in the players and performers who boldly color outside the lines: modders, hackers, and transgressive remixers, individuals who are easy to categorize as creative, original, and resistant to industry manipulation. Instead, I have chosen to investigate the rich possibilities of playing along: working within the constraints of game rules, commercial platforms, and existing repertoires. This has meant focusing on process rather than product. Recent digital-game design has embraced a similar orientation; gamemakers have been moving toward "research and design methods that help us create experiences that are less end-result based and more based in the quality of the experience itself—experiences with intrinsically motivating qualities" (Isbister, Flanagan et al. 2010:2041–2042). All competent *Guitar Hero* players who play a song are cueing playback of the same recording of that song, but that doesn't mean they all *feel* the same way every time—nor does it mean that producing that particular sonic output is what makes the experience satisfying.

Musical experience offers a powerful model for process-oriented game designers and analysts, in part because it is so much like play. It is immersive, unfolds in time, and is "intrinsically motivating." John Blacking's elegant working definition of music as "humanly organized sound" is seductive in its simplicity, but, as Blacking's own work demonstrates, music also depends on humanly organized time, bodies, and multisensory perception. (This is why playing *Guitar Hero* can feel as musical as playing a guitar.) All this "organization" is accomplished not just through flashes of individual creative inspiration but through technique and repertoire, rehearsal and repetition—through playing along. In popular-music and media studies, repetition often has negative connotations: of mass production, commodification, lack of originality, mind-numbing sameness consumed by a docile public. But scholars and practitioners of traditional rituals, games, and performing arts have a different relationship with repetition. In these repeating practices, each iteration may reinforce precedents or subtly alter them, gradually creating new traditional narratives, musical canons, embodied performance techniques, and cultural ideologies. Playing along with mainstream popular media is like performing any established repertoire: It entails repetition with a difference. Over time, those accumulated differences are transforming our understanding of musicality, creativity, play, and participation.

Playing along means accepting some structural constraints and embracing the productive potential of repetition. In this way players can frame a secure space for exploring new modes of experience, often in the supportive company of a community of practice. In the case studies presented here, playing along has yielded in-between experiences that destabilize conventional distinctions—including those dividing the avatar's body and the player's; listening and musicianship; production and reception; student and teacher; amateur and expert; conformity and creativity; compliance and resistance; immersion and detachment; the virtual and the actual. As interactive digital media increasingly make playing along a massively multiplayer affair, it will become ever more important to understand the compelling appeal and transformative potential of these pursuits.

NOTES

INTRODUCTION: PLAYING ALONG

1. These figures are drawn from NAMM 2009 and ESA 2010. Roberts, Foehr et al. 2010 offers a detailed breakdown of youth media consumption practices.
2. Ethnographic work has been rare in the digital-game studies literature. Important exceptions to the rule include T.L. Taylor's *Play Between Worlds* (2006)—a pioneering ethnography of massively multiplayer online games—and the ethnographically informed articles on *Dance Dance Revolution* by media scholar Jacob Smith (2004) and musicologist Joanna Demers (2006).
3. Willis's *Common Culture* (1990) set an important precedent for this book, but Willis and his collaborators conducted their research at a time when "the new screen-based technologies" were not yet widely available or used. They rightly noted that "there is a huge and yet unrealized potential here for the extension and development of young people's symbolic work and creativity" (42). The studies collected in Ito, Baumer et al. (2010a) make progress toward addressing the dearth of ethnographic work on digital media and participatory culture while also demonstrating how much remains to be done. For example, this ambitious and wide-ranging book about youth media practices only rarely addresses music.
4. See Lysloff (2003) and Boellstorff (2008) for responses to these concerns about "virtual" vs. "real" research associates.
5. Salen and Zimmerman's game design textbook *Rules of Play* helped popularize the "magic circle" metaphor among digital game designers and scholars. Salen and Zimmerman's own use of the term is complex; they acknowledge that the circle can be considered "open" under some circumstances (2004:96, 574). My argument is that considering it "closed" always distorts the nature of gameplay.
6. I use "perceived affordances" in the sense popularized by Donald Norman (1988, building on Gibson 1977). For historical accounts of this period, see, e.g., Chanan (1995); Katz (2004); Kenney (1999); Sterne (2003); Taylor (2001, 2005, 2007). For more on "liveness," see Auslander (1999) and Dixon (2007:115–134).
7. This sales data is compiled from figures cited in Berardini (2009), Graft (2009), Howe (2009), and Thorsen (2009).
8. The term Web 2.0 was popularized in connection with an Internet business conference hosted by O'Reilly Media in 2004 (and annually since then). See O'Reilly (2005).

CHAPTER 1: STRAIGHT OUTTA GANTON: VIRTUAL TOURISM, FIELDWORK, AND PERFORMANCE

1. I use italics to designate the digital game *San Andreas*—an authored multimedia text—and plain-text San Andreas to refer to the virtual place that constitutes the gameworld.

2. My approach owes a debt to the essays collected in Lysloff and Gay (2003).

3. See Lysloff (2003) for a reflexive, ethically astute account of web-based ethnography. Bell (2001) provides an overview of the first major writings on this topic; Boellstorff (2008) offers a more recent literature review and a strong defense of fieldwork in virtual worlds.

4. Unless otherwise noted, all quotations from game audio are transcribed from recorded gameplay.

5. Since poor inner-city African Americans are a minority of the U.S. population and rates of video game play are relatively consistent across race and class lines within that population, it is clear that poor inner-city African Americans do not represent "most" video game players. However, this should not be taken to indicate that the typical video game player is white and affluent. A 2005 study of in-home media availability found that 87 percent of African American children had access to a video game console at home, as compared to 82 percent of whites and Hispanics. Moreover, the lowest levels of parent education and household income correlated with the highest percentage of video game console availability (85–86 percent) (Roberts, Foehr et al. 2005:110). The 2010 update to this study confirmed that any digital divide across race and class lines was small and narrowing (Roberts, Foehr et al. 2010:23).

6. All survey respondents indicated how they wished to be cited (by name, geographic location, anonymously, etc.). I have cited players according to their preferences and have also included each player's age bracket.

7. *GTA IV*'s online multiplayer possibilities may be taking the series in another direction, but in my estimation they remain peripheral to the overall GTA gameplay experience.

8. I contacted some Rockstar Games personnel, but they declined to answer my questions on the record (apparently due to the logistical hurdle of obtaining company clearance). I therefore rely on published interviews when addressing the designers' perspective in the GTA chapters.

9. David Leonard has made this argument with respect to the *Tony Hawk* games, which "offer players the opportunity to play or dominate (black) city spaces in the absence of people of color" by creating virtual skate parks "where rules and laws do not pose threats or consequences to white men" (2005:112, 126).

10. Rockstar subsequently financed the production of *Sunday Driver* (2006), a documentary film about a Compton/Watts lowrider club.

11. The main gang colors in San Andreas are green, orange, and purple, a design decision widely regarded as a deliberate (and prudent) avoidance of the red and blue of L.A.'s Bloods and Crips.

12. In this transcription I follow the lead of the game's subtitles in using the spelling "holmes" rather than "homes" (i.e., "homeboy").

13. This summary analysis is based on the reviews collected in GameStats.com (2005). As one gaming magazine writer noted, "The real problem is not how [*San Andreas*] depicts the lead character, but that it's practically the only game with a black lead. Combined with how Rockstar depicts that lead, the picture of African Americans painted by video gaming looks bleak" (Wilburn 2005).

14. Transcribed from television coverage.
15. See Miller (2008a) for a detailed discussion of *GTA: San Andreas* as digital folk-lore. See Lhamon (1998:56–115) for a theory of the "blackface lore cycle."

CHAPTER 2: JACKING THE DIAL: RADIO, RACE, AND PLACE IN SAN ANDREAS

1. Scientist (Hopeton Browne) subsequently sued his label, Greensleeves Records, for licensing this material, asserting that he owned the copyright for the tracks in question. Greensleeves argued that Scientist was simply a mixing engineer and that the rights belonged to the producer (Marshall 2005).
2. Recorded interview, May 2006.
3. However, one metal-loving survey respondent did notice that in a cut-scene Tommy identifies himself as a fan of Lovefist, the game's fictional metal band; the player reported saying to himself "Alright, he's a metalhead" and feeling a closer connection with Tommy thereafter (Ben, Ontario, Canada).
4. Recorded interview, May 2006.
5. Recorded interview, May 2006.
6. The dub/reggae station suggests a Black Atlantic connection (Gilroy 1993); it seems more typical of Britain's urban black culture than of California's.
7. Apart from the evidence provided by my own ears, my main sources for these connections were the databases at www.the-breaks.com, www.secondhand-songs.com, and www.ishkur.com/samples.
8. "Straight Outta Compton" itself does not appear in *San Andreas*, but N.W.A.'s "Express Yourself" comes from the *Straight Outta Compton* album (1988); the album is also advertised in the game's instruction booklet. Narrative elements and locale aside, the visual style of *San Andreas* and the practice of indicating gang territory through colored outlines on maps have important parallels in the "Straight Outta Compton" video (available on YouTube via harvardpark617 2005). Gangsta rap scholar Eithne Quinn has analyzed this video and discussed the commodification of Compton as an index of gangsta authenticity (2005:77).
9. Nine tracks give primary artist credit to N.W.A., Dr. Dre, Ice Cube, the D.O.C., or Eazy-E; in addition, Dr. Dre and Ice Cube are credited as producers on the tracks by Above the Law and Da Lench Mob, respectively.
10. Several of my survey respondents mentioned that songs on the San Andreas dial are common sample sources. A German man wrote, "I noticed quite a few songs in San Andreas are the origin of samples used in modern pop songs, like Prodigy's Out Of Space [which samples "Critical Beatdown" by the Ultramagnetic MCs, on Playback FM], or a Saxophone sample used by Jennifer Lopez recently" (anonymous male, 22–26). The latter is presumably the saxophone sample from "Soul Power 74" (by Maceo & the Macks, on Master Sounds) featured in the 2005 J.Lo single "Get Right."
11. Some American soldiers serving in Iraq employed similar strategies in crafting a musical soundtrack for their work (Pieslak 2009). Pieslak's book makes almost no mention of video games. Perhaps a future study will investigate whether soldiers' ample experience with digital gameplay influences their musical practices in war zones (and vice versa).
12. Readers can view Madd Dogg's mansion in a YouTube contributor's machinima satire of MTV's celebrity lifestyle show *Cribs* (princearsalan0001 2006).
13. The Tascam DA-88 could be considered slightly anachronistic for the game-world, since it went to market in 1993. Alternatively, it could serve to mark Madd Dogg's equipment as cutting-edge.

14. Wheeler uses these terms to describe the 1982 Grandmaster Flash classic "The Message," which appears on *Vice City's* pirate rap station and is remixed as Ice Cube's "Check Yo Self" on Radio Los Santos.
15. This phrase was cited in a widely circulated AP story about a 2006 lawsuit filed against Rockstar Games and its parent companies in connection with 14-year-old Cody Posey's 2004 shooting of several family members in New Mexico (Associated Press 2006).

CHAPTER 3: HOW MUSICAL IS *GUITAR HERO*?

1. Dance pads are electronic floor mats that allow players to send input to the games using their feet; see Smith (2004) and Demers (2006).
2. I learned about *Guitar Hero's* backstory from an interview with Rob Kay, the game's lead designer, conducted in Cambridge, Massachusetts, on October 17, 2007. See also the interview with Kay published in Simons (2007).
3. With the release of *Rock Band 3* in late 2010, a keyboard controller joined the lineup.
4. This sales information is compiled from figures cited in Berardini (2009), Graft (2009), Howe (2009), Thorsen (2009), and Viacom (2010).
5. Recorded interview, July 9, 2009, in Boston, Massachusetts.
6. All survey respondents are quoted with permission and cited according to their stated preferences; each respondent's gender and age bracket is included.
7. Steffen, post-gameplay recorded interview, August 6, 2008.
8. Kevin, post-gameplay recorded interview, July 22, 2008.
9. The gender and age figures in particular probably say much more about the typical demographics of gamer messageboard participants than of *Guitar Hero* and *Rock Band* players; as I discuss later, these games were designed specifically to appeal to both male and female players.
10. Because I did not obtain Institutional Review Board approval to work with minors for this study, the lowest age bracket available on the survey form was 18–21. It is still likely that some respondents were under 18, since there was no mechanism for verifying respondents' stated age.
11. Roughly one-third of the respondents completed the survey before the first *Rock Band* game was released; it is very likely that many of them later played or purchased a *Rock Band* game.
12. Recorded interview, July 2, 2009, in Boston, Massachusetts.
13. Recorded interview, July 2, 2009 in Boston, Massachusetts.
14. Steffen, post-gameplay recorded interview, August 5, 2008.
15. Kevin, post-gameplay recorded interview, July 22, 2008.
16. Dustin, post-gameplay recorded interview, August 7, 2008.
17. Mike Dadmun, recorded interview, July 2, 2009, in Boston, Massachusetts.
18. Dan, post-gameplay recorded interview, August 6, 2008.
19. Mike, post-gameplay recorded interview, August 7, 2008.
20. Lauren, post-gameplay recorded interview, July 18, 2008.
21. Recorded interview, July 29, 2009, in Boston, Massachusetts.
22. Josh, post-gameplay recorded interview, July 29, 2008.
23. Colin, post-gameplay recorded interview, July 2, 2008.
24. Heather, recorded interview, July 2, 2009, in Boston, Massachusetts.
25. Adam, post-gameplay recorded interview, August 6, 2008.
26. Dan, post-gameplay recorded interview, August 6, 2008.
27. Kevin, post-gameplay recorded interview, July 22, 2008.

28. The main restriction on the available repertoire is the bottleneck created by the labor-intensive process of transcribing, mixing, and coding songs for release in *Rock Band*'s or *Guitar Hero*'s proprietary formats. In 2010 Harmonix introduced the "Rock Band Network," which was intended to address this bottleneck by allowing third parties to use company-supplied programming tools "to basically author up songs to our spec, submit them to a peer review process, and if they pass muster technically, put them on sale in the Rock Band store, for which we'll give them a respectable royalty" (DeRogatis and Kot 2009).

29. For more on flow and ritual, see Turner and Turner (1978:137). For discussions of immersion and flow in video games, see Ermi and Mäyrä (2005), Collins (2008:133), and Shultz (2008:178). For a critical reading of Csikszentmihalyi, see Sutton-Smith (1997:184–186).

30. Amandeep, post-gameplay recorded interview, July 23, 2008.

31. Kevin, post-gameplay recorded interview, July 22, 2008.

32. Recorded interview, July 9, 2009, in Boston, Massachusetts.

33. Steffen, post-gameplay recorded interview, August 5, 2008.

34. Sean, post-gameplay recorded interview, August 6, 2008.

35. Brian Shandra, recorded interview, July 9, 2009, in Boston, Massachusetts.

36. Ibid.

37. Recorded interview, October 17, 2007, in Cambridge, Massachusetts. All subsequent Rob Kay quotations are from this interview.

38. Recorded interview, July 9, 2009 in Boston, Massachusetts.

39. Kevin, post-gameplay recorded interview, July 22, 2008.

CHAPTER 4: JUST ADD PERFORMANCE: STAGING SCHIZOPHONIA

1. Because viewcounts wield so much power in search rankings, they are a contentious subject among YouTube uploaders. YouTube's online help site states that the company "employs proprietary technology to prevent the artificial inflation of a video's viewcount by spam bots, malware, and other means. We validate views to ensure the accuracy of the viewcount of all videos beginning with the first view. This validation process becomes publicly visible when the viewcount reaches 300. At this point, the viewcount may slow or temporarily freeze until we have time to verify that all further views are legitimate" (YouTube Help 2011; see also YouTube Help Forum 2009). YouTube does not reveal how much of a video someone has to watch before the view "counts," nor how it deals with validating repeat views from a single IP address.

2. "Gems" are the notes in the games' notation system; players get points for each note that they play correctly. The "star power" high-scoring mode, which multiplies the number of points one can earn for each note, is time-limited. Score-oriented players therefore make strategic choices about how to maximize earnable points within any given "star power" period, e.g., by saving "star power" for parts of a song that have a lot of notes.

3. Recorded interview, October 17, 2007, in Cambridge, Massachusetts.

4. Recorded interview, July 2, 2009, in Boston, Massachusetts.

5. Recorded interview, October 17, 2007, in Cambridge, Massachusetts.

6. The game software awards singers points based on how well their pitch and rhythm match up with the original vocal track. Thus a player who strays from the model provided by the original singer will get a low score, even if she or he has excellent vocal skills, sings in tune, etc.

7. Recorded interview, July 9, 2009, in Boston, Massachusetts.

8. Recorded interview, July 2, 2009, in Boston, Massachusetts.
9. In Wong's case, racial stereotyping also comes into play; numerous commenters link his Asianness with nerdishness or effeminacy.
10. Rob Drew makes a similar observation about karaoke: "It's hard not to feel that there's something excessive and off kilter at the heart of karaoke—a difference that can't be assimilated, a desire that can't be sated" (2001:124).
11. Recorded interview, July 29, 2009, in Boston, Massachusetts.
12. This YouTube video (freddiew 2008a) is no longer available; see MTV Networks (2008) for footage of the performance.
13. "Note streaks" and "FCs" are in-game scoring elements; the former represents the number of consecutive correct notes, and the latter stands for "full combo," the term that refers to playing an entire song with perfect note accuracy and no overstrumming (i.e., strumming when there are no notes). Both *Guitar Hero* and *Rock Band* present several different forms of evaluation at the end of each song, including a score, an accuracy percentage, and a star rating (up to five stars).

CHAPTER 5: MUSIC LESSONS 2.0

1. Unless otherwise noted, all David Taub quotes are from two recorded interviews, one conducted on the phone (November 10, 2009) and one in Taub's home studio in California (January 18, 2010).
2. Unless otherwise noted, all Nate Brown quotes are from a recorded phone interview on December 16, 2009.
3. Additional teachers are discussed in Chapter Six. I also interviewed a few teachers who offer one-on-one, real-time lessons via online video chat. This is a significantly different pedagogical format, but it shares some qualities with pre-recorded online lessons intended for a wide audience.
4. Recorded phone interview, February 27, 2010.
5. Ibid.
6. In the ensuing web forum discussion, Tim Gilberg explained that "the reason the picture is in the middle is that is where YouTube takes a snapshot of the image" to use as the video thumbnail (Next Level Guitar 2007a).
7. Email interview, August 5, 2010.
8. See U.S. Code Title 17, Section 107 (http://www.copyright.gov/title17/circ92.pdf). Fair use doctrine does specify that educational uses be "nonprofit," which could be problematic in the case of DrumMusic.tv.
9. Recorded phone interview, February 27, 2010.
10. This approach also reflects Nate Brown's training as a classroom teacher; in the course of pursuing his education degree he learned about the importance of teaching to a range of "learning styles."

CHAPTER 6: AMATEUR-TO-AMATEUR

1. F. Gregory Lastowka and Dan Hunter have used the term "amateur-to-amateur" to discuss peer-to-peer information practices and copyright law. For them, the term "amateur" highlights the fact that the "producer-participants" who engage in P2P practices "lack financial and proprietary motives" (Lastowka and Hunter 2006:3). Lastowka and Hunter use "amateur-to-amateur" interchangeably with "peer-to-peer," whereas I suggest that "amateur-to-amateur" usefully describes a specific subset of peer-to-peer practices.
2. All quotes from Nate Torres are from a recorded phone interview on December 4, 2009.

3. I regularly observe this authenticity effect in the classroom when I show YouTube videos to demonstrate basic techniques on various instruments. My students never question the authority of musicians whose ethnicity seems to "match" the musical tradition in question, but "mismatched" players often elicit snickers, or questions about how and why this person would have learned that instrument. Cf. Solís (2004).
4. In 2010, YouTube increased the limit to 15 minutes.
5. All quotes from John Dean are from an email interview completed on June 28, 2010.
6. Personal correspondence, reprinted with permission.
7. I contacted Grimmly to ask permission to feature his blog in this book; he graciously agreed. Because his blog already offers a wealth of information about his yoga practice, I opted not to take up more of his time by interviewing him.
8. This YouTube video features Steven Green and various other people practicing yoga amid gorgeous scenery in India. Short clips from different practice sessions are edited together into a single, flowing sequence.
9. See Keil (2010) for an anatomically oriented discussion of the *bandha* concept.
10. See Miller (2008b) for examples of similar discourse unfolding on Sacred Harp singing listservs.

ENDGAME
1. Recorded phone interview with Steve Wright, February 3, 2010.

REFERENCES

Aarseth, Espen J. 1997. *Cybertext: Perspectives on Ergodic Literature*. Baltimore, MD: Johns Hopkins University Press.

Adams, David. 2005. "Take-Two Financials Soar in Q1" (March 3, 2005). Accessed June 26, 2006. http://ps2.ign.com/articles/593/593230p1.html

Adorno, Theodor W. 1941. "On Popular Music." *Studies in Philosophy and Social Science* 9(1):17–48.

———. 1990 [1969]. "Opera and the Long-Playing Record." *October* 55:62–66.

———. 2001 [1938]. "On the Fetish Character in Music and the Regression of Listening." In *The Culture Industry: Selected Essays on Mass Culture*, ed. J. M. Bernstein. New York: Routledge.

Agawu, Kofi. 2003. *Representing African Music*. New York: Routledge.

Anderson, Benedict. 1991. *Imagined Communities: Reflections on the Origin and Spread of Nationalism*. New York: Verso.

Anderson, Kyle. 2007. "Even Better Than the Real Thing." *Spin* 23:104–110.

Arsenault, Dominic. 2008. *"Guitar Hero*: 'Not Like Playing Guitar at All'?" *Loading . . .* 2(2):n.p. http://journals.sfu.ca/loading/index.php/loading/article/view/32/29

Associated Press. 2006. "Lawsuit Blames Video Game for Posey Killings." Accessed February 21, 2007. http://www.abqjournal.com/news/apgrandtheft09-25-06.htm

Attali, Jacques. 1985 [1977]. *Noise: The Political Economy of Music*. Trans. Brian Massumi. Minneapolis: University of Minnesota Press.

Auslander, Philip. 1999. *Liveness: Performance in a Mediatized Culture*. New York: Routledge.

———. 2006. *Performing Glam Rock: Gender and Theatricality in Popular Music*. Ann Arbor: University of Michigan Press.

Babuscio, Jack. 1984. "Camp and the Gay Sensibility." In *Gays and Film*, ed. Richard Dyer. New York: Zoetrope.

backseatstuff. 2007. "Dueling Banjos Guitar Hero 2." (April 10, 2007). Accessed August 19, 2008. http://www.youtube.com/watch?v=8wfdgY5ZYDk

Barker, Martin and Kate Brooks. 1998. *Knowing Audiences: Judge Dredd, Its Friends, Fans, and Foes*. Luton, England: University of Luton Press.

Barthes, Roland. 1977 [1970]. "Musica Practica." Trans. Stephen Heath. In *Image—Music—Text*. New York: Hill and Wang.

Bartle, Richard. 1996. "Hearts, Clubs, Diamonds, Spades: Players Who Suit MUDs." *Journal of MUD Research* 1(1). Accessed December 15, 2006. http://www.mud.co.uk/richard/hcds.htm

Bausinger, Hermann. 1990. *Folk Culture in a World of Technology*. Trans. Elke Dettmer. Bloomington: Indiana University Press.

Bell, David. 2001. *An Introduction to Cybercultures*. New York: Routledge.

Berardini, César. 2009. "Rock Band Surpasses $1 Billion Dollars in Sales." *TeamXBox. com* (March 26, 2009). Accessed November 22, 2010. http://news.teamxbox. com/xbox/19228/Rock-Band-Surpasses-1-Billion-Dollars-in-Sales/

Berger, Harris M. 1999. *Metal, Rock, and Jazz: Perception and the Phenomenology of Musical Experience*. Hanover, NH: University Press of New England.

———. 2009. *Stance: Ideas about Emotion, Style, and Meaning for the Study of Expressive Culture*. Middletown, CT: Wesleyan University Press.

Berland, Jody. 1994. "Radio Space and Industrial Time: The Case of Music Formats." In *Canadian Music: Issues of Hegemony and Identity*, eds. Beverly Diamond and Robert Witmer. Toronto: Canadian Scholars' Press.

Blacking, John. 1973. *How Musical Is Man?* Seattle: University of Washington Press.

———. 1995. *Music, Culture, and Experience: Selected Papers of John Blacking*. Chicago: University of Chicago Press.

Blaine, Tina. 2005. "The Convergence of Alternate Controllers and Musical Interfaces in Interactive Entertainment." *Proceedings of the 2005 Conference on New Interfaces for Musical Expression (NIME)*: 27–33. Accessed November 9, 2010. http:// nime.org/2005/proc/nime2005_027.pdf

Blaine, Tina and Sidney Fels. 2003. "Contexts of Collaborative Musical Experiences." *Proceedings of the 2003 Conference on New Interfaces for Musical Expression (NIME)*: 129–134. Accessed November 9, 2010. http://www.nime.org/2003/ onlineproceedings/Papers/NIME03_Blaine.pdf

Boellstorff, Tom. 2008. *Coming of Age in Second Life: An Anthropologist Explores the Virtually Human*. Princeton, NJ: Princeton University Press.

Bourdieu, Pierre. 1977 [1972]. *Outline of a Theory of Practice*. Trans. Richard Nice. Cambridge: Cambridge University Press.

Bramwell, Tom. 2004. "Preview: Grand Theft Auto: San Andreas" (May 14, 2004). Accessed June 26, 2006. http://www.eurogamer.net/article.php?article_ id=55550

Britton, Stephen. 2004 [1991]. "Tourism, Capital and Place: Towards a Critical Geography of Tourism." In *Tourism: Critical Concepts in the Social Sciences*, ed. Stephen Williams. New York: Routledge.

Brocky1213. 2009. "Drum Tourney Hellashes Panic Attack SO CLOSE GameUnicon 2009." (September 1, 2009). Accessed November 3, 2010. http://www.youtube. com/watch?v=V6wvyh5FItw

Brownstein, Carrie. 2007. "Rock Band vs. Real Band." *Slate* (November 27, 2007). Accessed March 28, 2008. http://www.slate.com/id/2177432

Brunner, Cornelia. 2008. "Games and Technological Desire: Another Decade." In *Beyond Barbie and Mortal Kombat: New Perspectives on Gender and Gaming*, eds. Yasmin B. Kafai, Carrie Heeter, Jill Denner, and Jennifer Y. Sun. Cambridge, MA: MIT Press.

Burgess, Jean and Joshua Green. 2009. *YouTube: Online Video and Participatory Culture*. Cambridge: Polity Press.

Burrill, Derek A. 2006. "Check Out My Moves." *Social Semiotics* 16(1):17–38.

Butler, Judith. 1990. *Gender Trouble: Feminism and the Subversion of Identity*. New York: Routledge.

Butler, Mark J. 2006. *Unlocking the Groove: Rhythm, Meter, and Musical Design in Electronic Dance Music*. Bloomington: Indiana University Press.

Campbell, Patricia Shehan. 2004. *Teaching Music Globally: Experiencing Music, Expressing Culture*. New York: Oxford University Press.

Castronova, Edward. 2007. *Exodus to the Virtual World*. New York: Palgrave Macmillan.

Cecire, Natalia. 2010. "Virtuality and Distance Learning." *Arcade* (June 25, 2010). Accessed December 22, 2010. http://arcade.stanford.edu/pick-real-guitar-musica-practica-20#comment-626

Chanan, Michael. 1994. *Musica Practica: The Social Practice of Western Music from Gregorian Chant to Postmodernism*. New York: Verso.

———. 1995. *Repeated Takes: A Short History of Recording and Its Effects on Music*. New York: Verso.

Chappelle, David. 2003. "Grand Theft Auto." *Chappelle's Show* (Episode 109). Comedy Central. Accessed March 5, 2007. http://www.comedycentral.com/motherload/index.jhtml?ml_video=24421

———. 2006. "Tupac Lyrics." *Chappelle's Show* (Episode 301). Comedy Central. Accessed March 5, 2007. http://www.comedycentral.com/motherload/index.jhtml?ml_video=71416

Chess, Shira. 2005. "Playing the Bad Guy: *Grand Theft Auto* in the Panopticon." In *Digital Gameplay: Essays on the Nexus of Game and Gamer*, ed. Nate Garrelts. Jefferson, NC: McFarland.

Christensen, Thomas. 1999. "Four-Hand Piano Transcription and Geographies of Nineteenth-Century Musical Reception." *Journal of the American Musicological Society* 52(2):255–298.

Christgau, Robert. 2004. "U.S. and Them: Are American Pop (and Semi-Pop) Still Exceptional? And By the Way, Does That Make Them Better?" In *This Is Pop: In Search of the Elusive at Experience Music Project*, ed. Eric Weisbard. Cambridge, MA: Harvard University Press.

Clifford, James. 1988. *The Predicament of Culture*. Cambridge, MA: Harvard University Press.

———. 1997. *Routes: Travel and Translation in the Late Twentieth Century*. Cambridge, MA: Harvard University Press.

Collins, Karen. 2008. *Game Sound: An Introduction to the History, Theory, and Practice of Video Game Music and Sound Design*. Cambridge, MA: MIT Press.

Cooley, Timothy J., Katherine Meizel and Nasir Syed. 2008. "Virtual Fieldwork: Three Case Studies." In *Shadows in the Field: New Perspectives for Fieldwork in Ethnomusicology (Second Edition)*, eds. Gregory Barz and Timothy J. Cooley. New York: Oxford University Press.

Cottrell, Stephen. 2004. *Professional Music-Making in London: Ethnography and Experience*. Burlington, VT: Ashgate.

Csikszentmihalyi, Mihaly. 1991. *Flow: The Psychology of Optimal Experience*. New York: Harper Perennial.

Dahlen, Chris. 2008. "Harmonix Music Systems." *A.V. Club* (July 18, 2008). Accessed August 19, 2008. http://www.avclub.com/content/interview/harmonix_music_systems

de Certeau, Michel. 1984 [1974]. *The Practice of Everyday Life*. Trans. Steven Rendall. Berkeley: University of California Press.

de Marinis, Marco. 1987. "Dramaturgy of the Spectator." *TDR (The Drama Review)* 31(2):100–114.

Dégh, Linda. 1979. "Grimm's 'Household Tales' and Its Place in the Household: The Social Relevance of a Controversial Classic." *Western Folklore* 38(2):83–103.

Demers, Joanna. 2003. "Sampling the 1970s in Hip-Hop." *Popular Music* 22(1):41–56.

———. 2006. "Dancing Machines: 'Dance Dance Revolution,' Cybernetic Dance, and Musical Taste." *Popular Music* 25(3):401–414.

DeRogatis, Jim and Greg Kot. 2009. Interview with Greg LoPiccolo. *Sound Opinions* (Show 201). October 2, 2009. Chicago Public Radio and American Public Media.

Design Museum. 2002. "Rockstar Games: Multimedia Designers." Accessed June 26, 2006. http://www.designmuseum.org/design/index.php?id=67

Desmond, Jane C. 1993–1994. "Embodying Difference: Issues in Dance and Cultural Studies." *Cultural Critique* 26:33–63.

Devitt, Rachel. 2006. "Girl on Girl: Fat Femmes, Bio-Queens, and Redefining Drag." In *Queering the Popular Pitch*, eds. Sheila Whiteley and Jennifer Rycenga. New York: Routledge.

di Leonardo, Micaela. 1998. *Exotics at Home: Anthropologies, Others, American Modernity.* Chicago: University of Chicago Press.

Dixon, Steve. 2007. *Digital Performance: A History of New Media in Theater, Dance, Performance Art, and Installation.* Cambridge, MA: MIT Press.

Dorson, Richard M. 1976. *Folklore and Fakelore: Essays toward a Discipline of Folk Studies.* Cambridge, MA: Harvard University Press.

Drew, Rob. 2001. *Karaoke Nights: An Ethnographic Rhapsody.* Lanham, MD: Rowman & Littlefield.

Ermi, Laura and Frans Mäyrä. 2005. "Fundamental Components of the Gameplay Experience: Analysing Immersion." *Proceedings of DiGRA 2005 Conference: Changing Views—Worlds in Play*: 1–14. Accessed November 9, 2010. http://www.digra.org/dl/db/06276.41516.pdf

ESA [Entertainment Software Association]. 2010. "Essential Facts about the Computer and Video Game Industry." Accessed March 31, 2011. http://www.theesa.com/facts/pdfs/ESA_Essential_Facts_2010.pdf

Evans-Pritchard, Deirdre. 1987. "The Portal Case: Authenticity, Tourism, Traditions, and the Law." *Journal of American Folklore* 100(397):287–296.

Fabian, Johannes. 1983. *Time and the Other: How Anthropology Makes Its Object.* New York: Columbia University Press.

FallenSeraph. 2007. "SG Paint Job! check it out." (February 8, 2007). Accessed August 19, 2008. http://www.scorehero.com/forum/viewtopic.php?t=6333

FARK.com. 2006. "The Worst Songs in Video Games." Accessed August 27, 2006. http://forums.fark.com/cgi/fark/comments.pl?IDLink=2255544

Fine, Gary Alan. 1983. *Shared Fantasy: Role-Playing Games as Social Worlds.* Chicago: University of Chicago Press.

Finnegan, Ruth. 1989. *The Hidden Musicians: Music Making in an English Town.* Cambridge: Cambridge University Press.

Fish, Stanley. 1980. *Is There a Text in This Class? The Authority of Interpretive Communities.* Cambridge, MA: Harvard University Press.

Fornäs, Johan. 1998. "Filling Voids along the Byway: Identification and Interpretation in the Swedish Forms of Karaoke." In *Karaoke Around the World: Global Technology, Local Singing*, eds. Toru Mitsui and Shuhei Hosokawa. New York: Routledge.

Foster, Susan Leigh. 1998. "Choreographies of Gender." *Signs* 24(1):1–33.

Frasca, Gonzalo. 1999. "Ludology Meets Narratology: Similitude and Differences between (Video)games and Narrative." Accessed December 27, 2006. http://www.ludology.org/articles/ludology.htm

———. 2003. "Sim Sin City: Some Thoughts about Grand Theft Auto 3." *Game Studies* 3(2):n.p. Accessed September 29, 2005. http://www.gamestudies.org/0302/frasca

Fraser, Tara. 2006. *Total Astanga.* London: Duncan Baird.

freddiew. 2006. "Guitar Hero 2 Rush YYZ on Expert." (October 28, 2006). Accessed November 3, 2010. http://www.youtube.com/watch?v=Ua3hZXfNZOE

———. 2008a. "Hellanor Brozevelt on MTV TRL." (April 8, 2008). Accessed August 14, 2008. http://www.youtube.com/watch?v=yJSFVCODc6o

———. 2008b. "True Guitar Heroism." (April 17, 2008). Accessed November 3, 2010. http://www.youtube.com/watch?v=UFVyQSmf_RY

Frith, Simon. 1996. *Performing Rites: On the Value of Popular Music*. Cambridge, MA: Harvard University Press.

Fuller, Mary and Henry Jenkins. 1994. "Nintendo and New World Travel Writing: A Dialogue." In *CyberSociety: Computer-Mediated Communication and Community*, ed. Steven G. Jones. Thousand Oaks, CA: Sage.

GameStats.com. 2005. "Grand Theft Auto: San Andreas Articles." Accessed June 13, 2006. http://www.gamestats.com/objects/611/611957/articles.html#reviews

GameUniverse. 2009. "GameUniCon." Accessed November 3, 2010. http://www.scorehero.com/forum/viewtopic.php?t=90466

Gates, Henry Louis, Jr. 1988. *The Signifying Monkey: A Theory of African-American Literary Criticism*. New York: Oxford University Press.

Gauntlett, David. 2001. "The Worrying Influence of 'Media Effects' Studies." In *Ill Effects: The Media/Violence Debate (Second Edition)*, eds. Martin Barker and Julian Petley. New York: Routledge.

Gay, Leslie C., Jr. 1998. "Acting Up, Talking Tech: New York Rock Musicians and Their Metaphors of Technology." *Ethnomusicology* 42(1):81–98.

Gee, James Paul. 2004. *Situated Language and Learning*. New York: Routledge.

———. 2006. "Learning by Design: Good Video Games as Learning Machines." In *Digital Media: Transformations in Human Communication*, eds. Paul Messaris and Lee Humphreys. New York: Peter Lang.

Geertz, Clifford. 1973. *The Interpretation of Culture*. New York: Basic Books.

———. 1980. "Blurred Genres: The Refiguration of Social Thought." *American Scholar* 29(2):165–179.

Gibson, James J. 1977. "The Theory of Affordances." In *Perceiving, Acting, and Knowing: Toward an Ecological Psychology*, eds. Robert Shaw and John Bransford. Hillsdale, NJ: Lawrence Erlbaum Associates.

Gilroy, Paul. 1993. *The Black Atlantic: Modernity and Double Consciousness*. Cambridge, MA: Harvard University Press.

Glater, Jonathan D. 2008. "Hidden Sex Scenes Draw Ho-Hum, Except from Lawyers" (June 25, 2008). Accessed July 15, 2008. http://www.nytimes.com/2008/06/25/technology/25settle.html

Golub, Alex. 2010. "Being in the World (of Warcraft): Raiding, Realism, and Knowledge Production in a Massively Multiplayer Online Game." *Anthropological Quarterly* 83(1):17–46.

Gonzales, Amy L., Thomas Finley and Stuart Paul Duncan. 2009. "(Perceived) Interactivity: Does Interactivity Increase Enjoyment and Creative Identity in Artistic Spaces?" In *Proceedings of the 27th International Conference on Human Factors in Computing Systems, CHI 2009*, eds. Dan R. Olsen, Jr., Richard B. Arthur, Ken Hinckley, et al. Boston: Association for Computing Machinery.

Goodwin, Andrew. 1992. "Rationalization and Democratization in the New Technologies of Popular Music." In *Popular Music and Communication*, ed. James Lull. London: Sage.

Gordanius. 2007. "Poncho Sanchez—Fundimentals of Latin Music—Conga." (January 18, 2007). Accessed December 20, 2010. http://www.youtube.com/watch?v=SNQ3dLJKgro

Graburn, Nelson H. H. 1983. "The Anthropology of Tourism." *Annals of Tourism Research* 10(1):9–33.

Graft, Kris. 2009. "Rock Band: 10m Units Shipped Worldwide." *Edge-Online.com* (February 12, 2009). Accessed April 20, 2010. http://www.edge-online.com/news/rock-band-10m-units-shipped-worldwide

Grau, Oliver. 2003. *Virtual Art: From Illusion to Immersion*. Cambridge, MA: MIT Press.

grimmly2007. 2010a. "Ashtanga Exploring Dropbacks." (May 18, 2010). Accessed January 5, 2011. http://www.youtube.com/watch?v=jS9ceH1xssk

———. 2010b. "Assorted types of jump back and through." (June 8, 2010). Accessed April 21, 2011. http://www.youtube.com/watch?v=xBDPYqrQAgs

Grimmly. 2008a. "Jumping back." (July 9, 2008). Accessed December 20, 2010. http://grimmly2007.blogspot.com/2008/07/jumping-back.html

———. 2008b. "Owning my practice." (October 7, 2008). Accessed December 20, 2010. http://grimmly2007.blogspot.com/2008/10/owning-my-practice.html

———. 2008c. "When to start intermediate if your home practice." (December 8, 2008). Accessed December 20, 2010. http://grimmly2007.blogspot.com/2008/12/when-to-start-intermediate-if-your-home.html

———. 2009. "How to do a Straight leg Jump through blindfolded." (July 7, 2009). Accessed December 20, 2010. http://grimmly2007.blogspot.com/2009/06/how-to-do-straight-leg-jump-through.html

———. 2010. "Developing a home practice Parts 1–26." (May 16, 2010). Accessed December 20, 2010. http://grimmly2007.blogspot.com/2010/05/developing-home-practice-parts-1-26.html

———. 2011. "Back so soon?" (March 31, 2011). Accessed March 31, 2011. http://grimmly2007.blogspot.com/2011/03/back-so-soon.html

Grodal, Torben. 2003. "Stories for Eye, Ear, and Muscles: Video Games, Media, and Embodied Experience." In *The Video Game Theory Reader*, eds. Mark J. P. Wolf and Bernard Perron. New York: Routledge.

gtagaming.com. 2004. "Grand Theft Auto: San Andreas." Accessed September 21, 2010. http://www.gtagaming.com/forums/showthread.php?t=11853&page=5

Guinness World Records Gamer's Edition. 2008. "Chris Chike Reclaims His Guitar Hero III Guinness World Record." (September 5, 2008). Accessed October 5, 2010. http://gamers.guinnessworldrecords.com/news/050908_gh3_highscore.aspx

GuitarHeroTab.com. 2008. "GuitarHeroTab.com—Complete guitar tabs for Rock Band and Guitar Hero." Accessed August 20, 2008. http://guitarherotab.com

Gwinn, Eric. 2004. "If You Play 'San Andreas,' You'll Be a Black Male. Does It Matter?" *Chicago Tribune* (November 1, 2004). Accessed March 5, 2007. http://www.chicagotribune.com/technology/reviews/chi-0411010009nov01,1,5743024.story?page=1&ctrack=1&cset=true&coll=chi-technologyreviews-utl

Hahn, Tomie. 2007. *Sensational Knowledge: Embodying Culture Through Japanese Dance*. Middletown, CT: Wesleyan University Press.

Hamera, Judith. 2007. *Dancing Communities: Performance, Difference, and Connection in the Global City*. New York: Palgrave Macmillian.

Harmonix Music Systems. 2010a. "Harmonix Music Systems." Accessed October 12, 2010. http://www.harmonixmusic.com/#games

———. 2010b. "Rock Band 3 // Games // Rock Band." Accessed November 3, 2010. http://rockband.com/games/rb3

Hartlaub, Peter. 2007. "Rock Band, Guitar Hero III Video Game Do Rock, But Real Is Better." *San Francisco Chronicle* (November 27, 2007): E1. Accessed March 28, 2008. http://www.sfgate.com/cgi-bin/article.cgi?f=/c/a/2007/11/27/DDGNTHTE7.DTL&hw=hartlaub+guitar+hero&sn=001&sc=1000

harvardpark617. 2005. "Straight Outta Compton—NWA." Accessed September 20, 2010. http://www.youtube.com/watch?v=QkM_N5_otl4

Hermes, Joke. 2005. *Re-Reading Popular Culture*. Malden, MA: Blackwell.

Hiatt, Brian. 2008. "Secrets of the Guitar Heroes: John Mayer." *Rolling Stone* (June 12, 2008). Accessed August 19, 2008. http://www.rollingstone.com/news/story/21004549/secrets_of_the_guitar_heroes_john_mayer/2

Hill, Annette. 1997. *Shocking Entertainment: Viewer Response to Violent Movies*. Luton, England: University of Luton Press.

Hill, Logan and Dan Houser. 2008. "Rockstar Games' Dan Houser on Grand Theft Auto IV and Digitally Degentrifying New York" (May 2, 2008). Accessed July 15, 2008. http://nymag.com/daily/entertainment/2008/05/rockstar_games_dan_houser.html

Hills, Matt. 2002. *Fan Cultures*. New York: Routledge.

hooks, bell. 1994. *Outlaw Culture: Resisting Representations*. New York: Routledge.

———. 2004. *We Real Cool: Black Men and Masculinity*. New York: Routledge.

Horst, Heather, Becky Herr-Stephenson and Laura Robinson. 2010. "Media Ecologies." In *Hanging Out, Messing Around, and Geeking Out*, ed. Mizuko Ito. Cambridge, MA: MIT Press.

Houser, Sam. 2004. "Interview with Sam Houser." *GMR*. Accessed June 26, 2006. http://www.1up.com/do/feature?cId=3134519

Howe, Jeff. 2009. "Why the Music Industry Hates *Guitar Hero*." *Wired.com* (February 23, 2009). Accessed April 20, 2010. http://www.wired.com/culture/culturereviews/magazine/17-03/st_essay

Huizinga, Johan. 1955. *Homo Ludens: A Study of the Play Element in Culture*. Boston: Beacon Press.

———. 2003 [1938]. "Nature and Significance of Play as a Cultural Phenomenon." In *Performance: Critical Concepts in Literary and Cultural Studies* (Vol. One), ed. Philip Auslander. New York: Routledge.

iamchris4life. 2008. "Through the Fire and Flames TTFAF FC 100% 987,786." (June 3, 2008). Accessed November 3, 2010. http://www.youtube.com/watch?v=S5GpRJItqjw

Ice Cube and Angela Y. Davis. 1992. "Nappy Happy: A Conversation with Ice Cube and Angela Y. Davis." *Transition* 58:174–192.

igrandtheftauto.com. 2004. "What Does Your CJ Look Like?" Accessed September 21, 2010. http://www.igrandtheftauto.com/forums/index.php?showtopic=49147

Inspired Instruments. 2010. "You Rock Guitar | Learn Guitar | Innovative MIDI Guitar Technology." Accessed November 3, 2010. http://www.yourockguitar.com/

Isbister, Katherine, Mary Flanagan and Chelsea Hash. 2010. "Designing Games for Learning: Insights from Conversations with Designers." In *Proceedings of the 28th International Conference on Human Factors in Computing Systems, CHI 2010*, eds. Elizabeth D. Mynatt, Don Schoner, Geraldine Fitzpatrick, et al. Atlanta: Association for Computing Machinery.

Isbister, Katherine and Kristina Höök. 2007. "Evaluating Affective Interactions." *International Journal of Human-Computer Studies* 65(4):273–274.

———. 2009. "On Being Supple: In Search of Rigor without Rigidity in Meeting New Design and Evaluation Challenges for HCI Practitioners." In *Conference on*

Human Factors in Computing Systems: Proceedings of ACM CHI 2009. Boston: Association for Computing Machinery.

Ito, Mizuko, Sonja Baumer, et al. 2010a. *Hanging Out, Messing Around, and Geeking Out*. Cambridge, MA: MIT Press.

———. 2010b. "Introduction." In *Hanging Out, Messing Around, and Geeking Out*. Cambridge, MA: MIT Press.

Itzkoff, Dave. 2008. "Rec-Room Wizard." *New York Times* (August 10, 2008). Accessed August 19, 2008. http://www.nytimes.com/2008/08/10/arts/television/10itzk.html

Iyer, Vijay. 2008. "On Improvisation, Temporality, and Embodied Experience." In *Sound Unbound: Sampling Digital Music and Culture*, ed. Paul D. Miller. Cambridge, MA: MIT Press.

James, Robert M. 1974. *One*. Audio recording. CTI 6043.

Jenkins, Henry. 1992. *Textual Poachers: Television Fans and Participatory Culture*. New York: Routledge.

———. 2006a. *Convergence Culture: Where Old and New Media Collide*. New York: New York University Press.

———. 2006b. *Fans, Bloggers, and Gamers: Exploring Participatory Culture*. New York: New York University Press.

Jennings, David. 2007. *Net, Blogs and Rock 'n' Roll: How Digital Discovery Works and What It Means for Consumers, Creators and Culture*. London: Nicholas Brealey.

John. 2010. "Everything that's wrong with pop culture in two photos." *theCHIVE.com* (February 24, 2010). Accessed April 20, 2010. http://thechive.com/2010/02/24/everything-thats-wrong-with-pop-culture-in-two-photos-2-photos/

Jones, Richard O. 2007. "The Bad News and Good News of Obsessive Video Games." *Black Voice News* (February 8, 2007). Accessed March 5, 2007. http://www.blackvoicenews.com/content/view/40464/4/

Jones, Simon and Paul Willis. 1990. "Music and Symbolic Creativity." In *Common Culture*. Boulder, CO: Westview Press.

Jones, Steve and Kevin Featherly. 2002. "Re-Viewing Rock Writing: Narratives of Popular Music Criticism." In *Pop Music and the Press*, ed. Steve Jones. Philadelphia: Temple University Press.

Juul, Jesper. 2005. *Half-Real: Video Games between Real Rules and Fictional Worlds*. Cambridge, MA: MIT Press.

Kaivalya. 2010. "Astanga." (August 12, 2010). Accessed December 20, 2010. http://reluctantashtangi.blogspot.com/2010/08/astanga_12.html

Kamenetsky, Christa. 1992. *The Brothers Grimm and Their Critics: Folktales and the Quest for Meaning*. Athens: Ohio University Press.

Kant, Immanuel. 1998 [1781]. *Critique of Pure Reason*. Trans. Paul Guyer and Allen W. Wood. Cambridge: Cambridge University Press.

Kapchan, Deborah A. 1995. "Performance." *Journal of American Folklore* 108(430): 479–508.

Katz, Mark. 2004. *Capturing Sound: How Technology Has Changed Music*. Berkeley: University of California Press.

Keen, Andrew. 2007. *The Cult of the Amateur: How Today's Internet Is Killing Our Culture*. New York: Doubleday.

Keil, Charles. 1984. "Music Mediated and Live in Japan." *Ethnomusicology* 28(1):91–96.

Keil, David. 2010. "Mulha Bandha anatomically speaking." Accessed November 3, 2010. http://www.yoganatomy.com/articlemulha.html

Kenney, William Howland. 1999. *Recorded Music in American Life: The Phonograph and Popular Memory, 1890–1945*. New York: Oxford University Press.

Khan, Javed I. and Adam Wierzbicki. 2008. "Guest Editors' Introduction: Foundation of Peer-to-Peer Computing." *Computer Communications* 31(2):187–189.

Kingsbury, Henry. 1988. *Music, Talent, and Performance*. Philadelphia: Temple University Press.

Kirshenblatt-Gimblett, Barbara. 1996. "The Electronic Vernacular." In *Connected: Engagements with Media*, ed. George E. Marcus. Chicago: University of Chicago Press.

———. 1998a. *Destination Culture: Tourism, Museums, and Heritage*. Berkeley: University of California Press.

———. 1998b. "Folklore's Crisis." *Journal of American Folklore* 111(441):281–327.

Kline, Stephen, Nick Dyer-Witheford and Greig de Peuter. 2003. *Digital Play: The Interaction of Technology, Culture, and Marketing*. Montreal & Kingston: McGill-Queen's University Press.

Konami Corporation. 1998. *Dance Dance Revolution*. Tokyo: Konami Corporation.

Lahti, Martti. 2003. "As We Become Machines: Corporealized Pleasures in Video Games." In *The Video Game Theory Reader*, eds. Mark J. P. Wolf and Bernard Perron. New York: Routledge.

Lange, Patricia G. and Mizuko Ito. 2010. "Creative Production." In *Hanging Out, Messing Around, and Geeking Out*, ed. Mizuko Ito. Cambridge, MA: MIT Press.

Langfitt, Frank. 2007a. "Learning Guitar for Free (for Now) on YouTube." *Morning Edition* (April 7, 2007). Accessed July 1, 2010. http://www.npr.org/templates/story/story.php?storyId=11778602

———. 2007b. "YouTube Guitar Lessons Pulled in Copyright Spat." *Morning Edition* (July 6, 2007). Accessed July 1, 2010. http://www.npr.org/templates/story/story.php?storyId=11778602

Lastowka, F. Gregory and Dan Hunter. 2006. "Amateur-to-Amateur: The Rise of a New Creative Culture." *Policy Analysis* 567.

LaVigne, Chris. 2005. "Bloody Well Done." (January 17, 2005). Accessed February 21, 2007. http://www.game-brains.com/archive/jan17_2005/gta_sanandreas.htm

Lee, Kwan Min, Namkee Park and Seung-A Jin. 2006. "Narrative and Interactivity in Computer Games." In *Playing Video Games: Motives, Responses, and Consequences*, eds. Peter Vorderer and Jennings Bryant. Mahwah, NJ: Lawrence Erlbaum Associates.

Leonard, David J. 2005. "To the White Extreme: Conquering Athletic Space, White Manhood, and Racing Virtual Reality." In *Digital Gameplay: Essays on the Nexus of Game and Gamer*, ed. Nate Garrelts. Jefferson, NC: McFarland & Co.

Lhamon, Jr., W. T. 1998. *Raising Cain: Blackface Performance from Jim Crow to Hip Hop*. Cambridge, MA: Harvard University Press.

Loftus, Tom. 2004. "'Grand Theft Auto' Back with a Vengeance." (November 5, 2004). Accessed June 19, 2006. http://www.msnbc.msn.com/id/6399463/

Lott, Eric. 1993. *Love and Theft: Blackface Minstrelsy and the American Working Class*. New York: Oxford University Press.

lrockwood. 2006. "Yoga Stretching Exercises for Stretching Flexibility (#6)." (June 10, 2008). Accessed October 5, 2010. http://www.youtube.com/watch?v=F_GXujI-fqA

Lucas, Kenny. 2004. "Review of Grand Theft Auto: San Andreas" (November 4, 2004). Accessed June 15, 2008 (via http://www.archive.org). http://www.gamezilla.com/review.aspx?review=8965

Luerssen, John D. 2009. "Jack White and Jimmy Page Condemn 'Guitar Hero'." *Spinner.com* (June 23, 2009). Accessed April 20, 2010. http://www.spinner.com/2009/06/23/jack-white-and-jimmy-page-condemn-guitar-hero/

Lum, Casey Man Kong. 1996. *In Search of a Voice: Karaoke and the Construction of Identity in Chinese America*. Mahwah, NJ: L. Erlbaum Associates.

Lynch, Kevin. 2008. "Review: Grand Theft Auto IV" (April 29, 2008). Accessed July 15, 2008. http://www.mirror.co.uk/news/more-news/technology-gaming/gamezone/2008/04/29/review-grand-theft-auto-iv-89520-20399274/

Lysloff, René T. A. 2003. "Musical Life in Softcity: An Internet Ethnography." In *Music and Technoculture*, eds. René T. A. Lysloff and Leslie C. Gay, Jr. Middletown, CT: Wesleyan University Press.

Lysloff, René T. A. and Leslie C. Gay, Jr., eds. 2003. *Music and Technoculture*. Middletown, CT: Wesleyan University Press.

MacCannell, Dean. 2004 [1973]. "Staged Authenticity: Arrangements of Social Space in Tourist Settings." In *Tourism: Critical Concepts in the Social Sciences*, ed. Stephen Williams. New York: Routledge.

Marcus, George E. 1999. "The Uses of Complicity in the Changing Mise-en-Scène of Anthropological Fieldwork." In *The Fate of Culture*, ed. Sherry Ortner. Berkeley: University of California Press.

Marriott, Michel. 2004. "The Color of Mayhem." (August 12, 2004). Accessed June 19, 2006. http://tech2.nytimes.com/mem/technology/techreview.html?res=9502E6DE163FF931A2575BC0A9629C8B63

Marshall, Wayne. 2005. "Mad Scientist, for Good Reason." Accessed September 20, 2010. http://wayneandwax.blogspot.com/search?q=scientist

McGranahan, Liam. 2010. "Mashnography: Creativity, Consumption, and Copyright in the Mashup Community." Ph.D. diss., Brown University.

Mechling, Jay. 1984. "Patois and Paradox in a Boy Scout Treasure Hunt." *Journal of American Folklore* 97(383):24–42.

Messaris, Paul and Lee Humphreys, eds. 2006. *Digital Media: Transformations in Human Communication*. New York: Peter Lang.

Meyer, Moe. 1994. "Introduction: Reclaiming the Discourse of Camp." In *The Politics and Poetics of Camp*. New York: Routledge.

Millard, André and Rebecca McSwain. 2004. "The Guitar Hero." In *The Electric Guitar: A History of an American Icon*, ed. André Millard. Baltimore: Johns Hopkins University Press.

Miller, Kiri. 2005. "Jacking the Dial: A Grand Theft Auto Radio Survey." Accessed April 20, 2010. http://www.brown.edu/Project/Music/gtasurvey.html

———. 2007a. "e-interview with Freddie Wong." Guitar Hero: A Research Blog (November 1, 2007). Accessed November 3, 2010. http://guitarheroresearch.blogspot.com/2007/11/e-interview-with-freddie-wong.html

———. 2007b. "Guitar Hero & Rock Band Survey." Accessed November 3, 2010. http://www.brown.edu/Project/Music/guitarherosurvey2.html

———. 2008a. "Grove Street Grimm: Grand Theft Auto and Digital Folklore." *Journal of American Folklore* 121(481):255–285.

———. 2008b. *Traveling Home: Sacred Harp Singing and American Pluralism*. Urbana: University of Illinois Press.

Miller, Leta E. 2001. "Cage, Cunningham, and Collaborators: The Odyssey of *Variations V*." *The Musical Quarterly* 85(3):545–567.

Miller, Matt. n.d. "Grand Theft Auto: San Andreas: Second Opinion." *Game Informer*. Accessed March 5, 2007. http://www.gameinformer.com/NR/exeres/51D890DA-15D6-40F7-AD75-EC36C6C770A7.htm

Mitsui, Toru and Shuhei Hosokawa, eds. 1998. *Karaoke Around the World: Global Technology, Local Singing.* New York: Routledge.

Moby Games. 2010. "Grand Theft Auto series." Accessed September 20, 2010. http://www.mobygames.com/game-group/grand-theft-auto-series

Morris, Sue. 2004. "Co-Creative Media: Online Multiplayer Computer Game Culture." *Scan* 1(1). Accessed December 26, 2006. http://scan.net.au/scan/journal/display.php?journal_id=16

Motz, Marilyn. 1998. "The Practice of Belief." *Journal of American Folklore* 111(441):339–355.

MTV Networks. 2008. "Hellanor Brozevelt Perform 'Blitzkrieg Bop'." Accessed November 3, 2010. http://www.mtv.com/videos/misc/189609/hellanor-brozevelt-perform-blitzkrieg-bop.jhtml

Munster, Anna. 2006. *Materializing New Media.* Hanover, NH: Dartmouth College Press.

Murray, 1997 Murray, Janet H. 1997. *Hamlet on the Holodeck: The Future of Narrative in Cyberspace.* New York: Free Press.

Murray, 2005 Murray, Soraya. 2005. "High Art/Low Life: The Art of Playing *Grand Theft Auto.*" *PAJ* 27(2):91–98.

MySpace Forums. 2008. "Gutair Hero." Accessed October 12, 2010. http://forums.myspace.com/t/4019948.aspx?fuseaction=forums.viewthread&PageIndex=10&SortOrder=1

Nader, Laura. 1972. "Up the Anthropologist—Perspectives Gained from Studying Up." In *Reinventing Anthropology*, ed. Dell H. Hymes. New York: Pantheon.

NAMM [National Association of Music Merchants]. 2009. "2009 NAMM Global Report: A Statistical Review of the Music Products Industry." Accessed March 31, 2011. http://www.nxtbook.com/nxtbooks/namm/2009musicusa/

Naqvi, Husain. 2005. "Critical Digressions: Gangbanging and Notions of the Self." *3 Quarks Daily* (August 29, 2005). Accessed March 5, 2007. http://3quarksdaily.blogs.com/3quarksdaily/2005/08/critical_digres_3.html

Nettl, Bruno. 1995. *Heartland Excursions: Ethnomusicological Reflections on Schools of Music.* Urbana: University of Illinois Press.

Next Level Guitar. 2007a. "Ditch the girls." *Next Level Guitar Forum* (July 23, 2007). Accessed December 22, 2010. http://www.nextlevelguitar.com/aforum/showthread.php?t=7478

———. 2007b. "Fellow Beginners, Show youselves!" *Next Level Guitar Forum* (January 27, 2007). Accessed December 22, 2010. http://www.nextlevelguitar.com/aforum/showthread.php?t=17

———. 2008. "Just learned my first real solo." *Next Level Guitar Forum* (September 2, 2008). Accessed December 22, 2010. http://www.nextlevelguitar.com/aforum/showthread.php?t=10745

Niles, Richard. 2004. "Wigs, Laughter, and Subversion: Charles Busch and Strategies of Drag Performance." *Journal of Homosexuality* 46(3–4):35–53.

Norman, Donald. 1988. *The Psychology of Everyday Things.* New York: Basic Books.

O'Reilly, Tim. 2005. "What Is Web 2.0?" Accessed December 4, 2010. http://oreilly.com/web2/archive/what-is-web-20.html

onlinedrummer. 2006. "Drum Lesson 'Lessons From Back in Black' by ACDC." (February 10, 2006). Accessed December 20, 2010. http://www.youtube.com/watch?v=nnfLZ0HsLiU

———. 2007. "Drum Lesson: Learning to Read 16th Note Triplets" (October 6, 2007). Accessed April 21, 2011. http://www.youtube.com/watch?v=P_ki8hIMH2M

———. 2010a. "Drum Lesson—A Funky Groove—Nate Brown." (April 16, 2010). Accessed January 5, 2011. http://www.youtube.com/watch?v=7Cl4vLi5aPA

———. 2010b. "Drum Lesson—My First Samba—Nate Brown." (May 23, 2010). Accessed April 21, 2011. http://www.youtube.com/watch?v=b3X3KBIXTwY

OnlineDrummer.com. 2004. "Drum Beat—Bellin." (December 13, 2004). Accessed December 20, 2010. http://onlinedrummer.com/video.php?BeatId=112

———. 2010a. "Drum Forum—Statistics Center." Accessed December 21, 2010. http://www.onlinedrummer.com/forum/index.php?action=stats

———. 2010b. "How do you 'just jam' ??" (February 20, 2010). Accessed December 22, 2010. http://www.onlinedrummer.com/forum/index.php?topic=29150.0

———. 2010c. "I HAVE AN AUDITION AT DRUM-TECH IN LONDON. I'm nervous." (February 11, 2010). Accessed December 22, 2010. http://www.onlinedrummer.com/forum/index.php?topic=29055.0

———. 2010d. "My First Samba." Accessed December 20, 2010. http://www.onlinedrummer.com/drum_lesson_video.php?Id=109

Owl. 2009. "DIY and Not-So-Private Minds." (May 10, 2009). Accessed December 22, 2010. http://www.insideowl.com/article/diy-and-not-so-private-minds

P4721cX. 2008. "The Ultimate GH/RB Fender SG." (March 4, 2008). Accessed August 19, 2008. http://www.scorehero.com/forum/viewtopic.php?t=54807

Parker, James. 2009. "School of Rock." *The Atlantic* (March 2009). Accessed March 24, 2009. http://www.theatlantic.com/doc/200903/guitar-hero

pianojohn113. 2007a. "Billy Joel #1—How to Play Piano Man Intro." (June 9, 2007). Accessed December 20, 2010. http://www.youtube.com/watch?v=U9Mhy1KmC04

———. 2007b. "Pianojohn113's Channel." (Update version: April 4, 2011). Accessed April 12, 2011. http://www.youtube.com/user/pianojohn113

———. 2010. "A Pianojohn113 Bulletin!" (March 27, 2010). Accessed December 20, 2010. http://www.youtube.com/watch?v=YwNJHah9ze8

Pieslak, Jonathan. 2009. *Sound Targets: American Soldiers and Music in the Iraq War.* Bloomington: Indiana University Press.

Potts, Jason, John Hartley, et al. 2008. "Consumer Co-Creation and Situated Creativity." *Industry and Innovation* 15(5):459–474.

princearsalan0001. 2006. "Grand Theft Auto CRIBS." (March 4, 2006). Accessed February 10, 2007. http://www.youtube.com/watch?v=amfXIIxhaD8

prpapito3000. 2007a. "Conga Lesson 1: Basic Tones—Nate Torres." (October 9, 2007). Accessed December 22, 2010. http://www.youtube.com/watch?v=3XLqe4eBSSk

———. 2007b. "Conga Lesson 4: Basic Tumbao/Rhythm for Salsa—Nate Torres." (October 9, 2007). Accessed December 22, 2010. http://www.youtube.com/watch?v=dvAqDWlTzpc

———. 2007c. "Conga Lesson 5: Basic Tumbao Variations Salsa—Nate Torres " (October 17, 2007). Accessed December 22, 2010. http://www.youtube.com/watch?v=VUf9s4z-TNQ

———. 2008a. "Conga Lesson 6: Salsa Tumbao for Two Congas—Nate Torres." (January 1, 2008). Accessed December 22, 2010. http://www.youtube.com/watch?v=DlaNeDL3Kvw

———. 2008b. "Conga Lesson 8: How to Play Rumba Guaguanco—Nate Torres." (June 25, 2008). Accessed December 22, 2010. http://www.youtube.com/watch?v=UOSABaDhRAk

PS3.QJ.net. 2007. "Writer Says GTA is Racist—Gamers React." Accessed March 5, 2007. http://ps3.qj.net/Writer-says-GTA-is-racist-gamers-react/pg/49/aid/83184

PS3Forums.com. 2004. "NY Times' issues with GTA San Andreas 'racial implicati[ons]'." Accessed March 5, 2007. http://www.ps3forums.com/showthread.php?t=420

Quinn, Eithne. 2005. *Nuthin' but a "G" Thang: The Culture and Commerce of Gangsta Rap*. New York: Columbia University Press.

Roberts, Donald F., Ulla G. Foehr and Victoria Rideout. 2005. *Generation M: Media in the Lives of 8–18 Year-Olds*. Menlo Park, CA: Henry J. Kaiser Family Foundation.

———. 2010. *Generation M2: Media in the Lives of 8–18 Year-Olds*. Menlo Park, CA: Henry J. Kaiser Family Foundation.

rockongoodpeople. 2007. "If you want the TRUTH Click the link in the info box." (July 3, 2007). Accessed December 22, 2010. http://www.youtube.com/watch?v=Yf4iy9uIO34

———. 2008. "Learning Guitar anyone can play, Q&A guitar hero talk." (April 26, 2008). Accessed March 31, 2011. http://www.youtube.com/watch?v=T_cQHTF_dY0

———. 2010. "Acoustic guitar lesson add licks to strum patterns combine parts song technique on Taylor GS5." (July 4, 2010). Accessed January 5, 2011. http://www.youtube.com/watch?v=nt56Ww86jLA

Rockstar Games. 2004a. *Grand Theft Auto: San Andreas*. New York: Take-Two Interactive.

———. 2004b. *Grand Theft Auto: San Andreas: Local Business Advertiser's Guide [instruction booklet]*. New York: Take-Two Interactive.

———. 2004c. "Grand Theft Auto San Andreas Screen 10." Accessed April 21, 2011. http://www.rockstargames.com/sanandreas/screens/screen10.html

———. 2004d. "Grand Theft Auto San Andreas Screen 11." Accessed April 21, 2011. http://www.rockstargames.com/sanandreas/screens/screen11.html

———. 2008a. *Grand Theft Auto IV*. New York: Take-Two Interactive.

———. 2008b. *Libery City Guidebook [GTA IV instruction booklet]*. New York: Take-Two Interactive.

Rose, Tricia. 1994a. *Black Noise: Rap Music and Black Culture in Contemporary America*. Middletown, CT: Wesleyan University Press.

———. 1994b. "A Style Nobody Can Deal With: Politics, Style and the Postindustrial City in Hip Hop." In *Microphone Fiends: Youth Music and Youth Culture*, eds. Andrew Ross and Tricia Rose. New York: Routledge.

Ross, Andrew. 1989. *No Respect: Intellectuals and Popular Culture*. New York: Routledge.

Ross, Michael. 2008. "The Real Heroes of Guitar Hero III." *Guitar Player* 42:58–63.

Ryan, Marie-Laure. 1999. "Cyberspace, Virtuality, and the Text." In *Cyberspace Textuality: Computer Technology and Literary Theory*, ed. Marie-Laure Ryan. Bloomington: Indiana University Press.

Salen, Katie and Eric Zimmerman. 2004. *Rules of Play: Game Design Fundamentals*. Cambridge, MA: MIT Press.

Salvato, Nick. 2009. "Out of Hand: YouTube Amateurs and Professionals." *TDR* 53(3):67–83.

Samudra, Jaida Kim. 2008. "Memory in Our Body: Thick Participation and the Translation of Kinesthetic Experience." *American Ethnologist* 35(4):665–681.

Schafer, R. Murray. 1969. *The New Soundscape: A Handbook for the Modern Music Teacher*. Don Mills, Ontario: BMI Canada Limited.

Schechner, Richard. 1985. *Between Theater and Anthropology*. Philadelphia: University of Pennsylvania Press.

Schloss, Joseph G. 2004. *Making Beats: The Art of Sample-Based Hip-Hop*. Middletown, CT: Wesleyan University Press.

Schutz, Alfred. 1964–67 [1951]. "Making Music Together: A Study in Social Relationship." In *Collected Papers, Volume II: Studies in Social History*, ed. Arvid Brodersen. The Hague: M. Nijhoff.

Scorehero.com. 2010. "Drum Modding 101." Accessed November 3, 2010. http://rockband.scorehero.com/forum/viewtopic.php?t=3030

Shandra, Brian. 2010. "Score Hero Wiki: User_BriGuy." Accessed November 3, 2010. http://wiki.scorehero.com/User_BriGuy

Shultz, Peter. 2008. "Music Theory in Music Games." In *From Pac-Man to Pop Music: Interactive Audio in Games and New Media*, ed. Karen Collins. Burlington, VT: Ashgate.

Shuman, Amy. 1986. *Storytelling Rights: The Uses of Oral and Written Texts by Urban Adolescents*. Cambridge: Cambridge University Press.

Simons, Iain. 2007. *Inside Game Design*. London: Laurence King.

Sinnreich, Aram Arthur. 2007. "Configurable Culture: Mainstreaming the Remix, Remixing the Mainstream." Ph.D. diss., University of Southern California.

SKPJAYI [Shri K Pattabhi Jois Ashtanga Yoga Institute]. 2009. *"Teachers Information."* Accessed December 22, 2010. http://kpjayi.org/the-institute/teachers

Small, Christopher. 1998. *Musicking: The Meanings of Performance and Listening*. Hanover, NH: University Press of New England.

Smith, Jacob. 2004. "I Can See Tomorrow in Your Dance: A Study of Dance Dance Revolution and Music Video Games." *Journal of Popular Music Studies* 16(1): 58–84.

Solís, Ted. 2004. *Performing Ethnomusicology: Teaching and Representation in World Music Ensembles*. Berkeley: University of California Press.

soulfulbohemian. 2008. "Ralph Tresvant—'Sensitivity'." (August 17, 2008). Accessed September 27, 2010. http://www.youtube.com/watch?v=c_28hZyxu5I

South Park Digital Studios. 2007. *Guitar Queer-o* (November 7, 2007). http://www.southparkstudios.com/episodes/127947

Spence D. 2004. "Grand Theft Auto: San Andreas—Radio Station Blowout." Accessed September 7, 2006. http://music.ign.com/articles/575/575710p1.html

Squire, Kurt. 2008. "Open-Ended Video Games: A Model for Developing Learning for the Interactive Age." In *The Ecology of Games: Connecting Youth, Games, and Learning*, ed. Katie Salen. Cambridge, MA: MIT Press.

Stallock, Kyle. 2008. "Teen breaks own world record in Guitar Hero III." *1Up.com* (July 4, 2008). Accessed August 19, 2008. http://www.1up.com/do/newsStory?cId=3168547

States, Bert O. 2003 [1996]. "Performance as Metaphor." In *Performance: Critical Concepts in Literary and Cultural Studies* (Vol. One), ed. Philip Auslander. New York: Routledge.

Stebbins, Robert A. 1992. *Amateurs, Professionals, and Serious Leisure*. Montreal: McGill-Queen's University Press.

Sterne, Jonathan. 2003. *The Audible Past: Cultural Origins of Sound Reproduction*. Durham, NC: Duke University Press.

Stokes, Martin. 1999. "Music, Travel, and Tourism: An Afterword." *World of Music* 41(3):141–155.

Strain, Ellen. 2003. *Public Places, Private Journeys: Ethnography, Entertainment, and the Tourist Gaze*. New Brunswick, NJ: Rutgers University Press.

Surette, Tim. 2004. "San Andreas Assaults UK Sales Record" (November 10, 2004). Accessed June 26, 2006. http://www.gamespot.com/ps2/action/gta4/news.html?sid=6112876

Sutton-Smith, Brian. 1983. "Play Theory and Cruel Play of the Nineteenth Century." In *The World of Play: Proceedings of the 7th Annual Meeting of the Association of the Anthropological Study of Play*, ed. Frank Manning. West Point, NY: Leisure Press.

———. 1995. "The Persuasive Rhetorics of Play." In *The Future of Play Theory*, ed. Anthony D. Pellegrini. Albany: State University of New York Press.

———. 1997. *The Ambiguity of Play*. Cambridge, MA: Harvard University Press.

Svec, Henry Adam. 2008. "Becoming Machinic Virtuosos: Guitar Hero, Rez, and Multitudinous Aesthetics." *Loading . . .* 2(2):n.p. http://journals.sfu.ca/loading/index.php/loading/article/view/30/28

Take-Two Interactive. 2008. "Recommendation of the Board of Directors to Reject Electronic Arts Inc.'s Tender Offer March 2008." Accessed December 22, 2010. http://web.archive.org/web/20080408234728/http://taketwovalue.com/documents/TTWO_Value.pdf#page=12

———. 2010. "Take-Two Interactive Software, Inc. Reports Second Quarter Fiscal 2010 Financial Results" (June 8, 2010). Accessed December 22, 2010. http://web.archive.org/web/20080408234728/http://taketwovalue.com/documents/TTWO_Value.pdf#page=12

Taylor, T. L. 2006. *Play Between Worlds: Exploring Online Game Culture*. Cambridge, MA: MIT Press.

Taylor, Timothy D. 2001. *Strange Sounds: Music, Technology and Culture*. New York: Routledge.

———. 2005. "Music and the Rise of the Radio in Twenties America: Technological Imperialism, Socialization, and the Transformation of Intimacy." In *Wired for Sound: Engineering and Technologies in Sonic Cultures*, eds. Paul D. Greene and Thomas Porcello. Middletown, CT: Wesleyan University Press.

———. 2006. "Music and Digital Culture: New Forms of Consumption and Commodification." In *Digital Media: Transformations in Human Communication*, eds. Paul Messaris and Lee Humphreys. New York: Peter Lang.

———. 2007. "The Commodification of Music at the Dawn of the Era of 'Mechanical Music'." *Ethnomusicology* 51(2):281–305.

Théberge, Paul. 1997. *Any Sound You Can Imagine: Making Music/Consuming Technology*. Middletown, CT: Wesleyan University Press.

TheYellowSubmarine13. 2009. "TheYellowSubmarine13's Channel." (June 22, 2009). Accessed December 22, 2010. http://www.youtube.com/user/TheYellowSubmarine13

———. 2010. "Real Love—John Lennon." (June 15, 2010). Accessed December 22, 2010. http://www.youtube.com/watch?v=n2_TXuNQdfU

Thompson, Clive. 2006. "Survey: What Music Do You Listen To While Playing Grand Theft Auto?" Accessed January 11, 2006. http://www.collisiondetection.net/mt/archives/2006/01/

Thornton, Sarah. 1996. *Club Cultures: Music, Media and Subcultural Capital*. Hanover, NH: Wesleyan University Press.

Thorsen, Tor. 2009. "Guitar Hero tops $2 billion, Activision Blizzard earns $981 million in Q1." *GameSpot.com* (May 7, 2009). Accessed November 22, 2010. http://www.gamespot.com/news/6209327.html

———. 2010. "NPD: Wii Play top US best-seller to date." *GameSpot.com* (January 19, 2010). Accessed December 21, 2010. http://uk.gamespot.com/news/6246627.html

Titon, Jeff Todd. 1995. "Text." *Journal of American Folklore* 108(430):432–448.

———. 1997. "Knowing Fieldwork." In *Shadows in the Field*, eds. Gregory F. Barz and Timothy J. Cooley. Oxford: Oxford University Press.

Turkle, Sherry. 2011. *Alone Together: Why We Expect More from Technology and Less from Each Other*. New York: Basic Books.

Turner, Victor. 1974. *Dramas, Fields, and Metaphors*. Ithaca, NY: Cornell University Press.

Turner, Victor and Edith Turner. 1978. *Image and Pilgrimage in Christian Culture*. New York: Columbia University Press.

Vargas, Jose Antonio. 2005. "Gamers' Intersection." *Washington Post* (September 27, 2005). Accessed March 5, 2007. http://www.washingtonpost.com/wp-dyn/content/article/2005/09/26/AR2005092601697.html

Veal, Michael E. 2007. *Dub: Soundscapes and Shattered Songs in Jamaican Reggae*. Middletown, CT: Wesleyan University Press.

Viacom. 2010. "Rock Band Music Catalog Surpasses 2,000-Song Milestone with the Jimi Hendrix Experience's Classic Song, 'Are You Experienced?'" (October 5, 2010). Accessed October 26, 2010. http://www.viacom.com/news/Pages/newstext.aspx?RID=1479151

Waksman, Steve. 1999. *Instruments of Desire: The Electric Guitar and the Shaping of Musical Experience*. Cambridge, MA: Harvard University Press.

———. 2001. "Into the Arena: Edward Van Halen and the Cultural Contradictions of the Guitar Hero." In *Guitar Cultures*, eds. Andy Bennett and Kevin Dawe. New York: Berg.

Walser, Robert. 1993. *Running with the Devil: Power, Gender, and Madness in Heavy Metal Music*. Hanover, NH: University Press of New England.

Walters, Ricky. 1988. *The Great Adventures of Slick Rick*. Audio recording. 40513. Def Jam Recordings.

Weber, Samuel. 2004. *Theatricality as Medium*. New York: Fordham University Press.

Weinstein, Deena. 2004. "Creativity and Band Dynamics." In *This Is Pop: In Search of the Elusive at Experience Music Project*, ed. Eric Weisbard. Cambridge, MA: Harvard University Press.

Whalen, Zach. 2004. "Play Along—An Approach to Videogame Music." *Game Studies* 4(1). Accessed September 29, 2005. http://www.gamestudies.org/0401/whalen/

Wheeler, Elizabeth A. 1991. "'Most of My Heroes Don't Appear on No Stamps': The Dialogics of Rap Music." *Black Music Research Journal* 11(2):193–216.

White, Hayden. 1980. "The Value of Narrativity in the Representation of Reality." *Critical Inquiry* 7(1):5–27.

Wilburn, Thomas. 2005. "Guns, Gangs, and Greed." *The Escapist*. Accessed March 5, 2007. http://www.escapistmagazine.com/print/15/4

Willis, Paul. 1990. *Common Culture*. Boulder, CO: Westview Press.

———. 2000. *The Ethnographic Imagination*. Malden, MA: Polity Press.

WNYC. 2010. "Online Music Lessons Smackdown." *Soundcheck* (May 18, 2010). Accessed December 20, 2010. http://beta.wnyc.org/shows/soundcheck/2010/may/18/smackdown-online-music-lessons/

wumpus. 2007. "Adding weights to the SG to give it 'heft'." (November 25, 2007). Accessed August 19, 2008. http://www.scorehero.com/forum/viewtopic.php?t=40694

Wurtzler, Steve. 1992. "'She Sang Live, but the Microphone Was Turned Off': The Live, the Recorded and the Subject of Representation." In *Sound Theory/Sound Practice*, ed. Rick Altman. New York: Routledge.

xkcd. n.d. "Rock Band." Accessed December 22, 2010. http://xkcd.com/359/

Young Maylay. 2005. *San Andreas: The Original Mixtape*. Studio City, CA: Maylaynium Muziq.

YouTube Help. 2011. "Why is my viewcount frozen?" Accessed April 12, 2011. http://www.google.com/support/youtube/bin/answer.py?hl=en&answer=175736

YouTube Help Forum. 2009. "How does the views count?" (June 10, 2009). Accessed April 12, 2011. http://www.google.com/support/forum/p/youtube/thread?tid=6abc6fdf32d2b278&hl=en

Zezima, Katie. 2007. "Virtual Frets, Actual Sweat." *New York Times* (July 15, 2007). Accessed November 3, 2010. http://www.nytimes.com/2007/07/15/fashion/15guitar.html

INDEX

paideia, 26

pedagogy. *See* online music pedagogy; online yoga pedagogy

peer-to-peer (P2P), 183–184, 225

performance, 4–5, 7, 44, 103, 120–122, 25–137, 139, 141–151, 223. *See also* theatricality

performativity, 16, 44, 129, 143

permeability, 86

Pianojohn113. *See* Dean, John

play
  theories of, 5, 10–11, 26, 33, 36, 49–50, 51. *See also* digital gameplay

playback technology and musical practice, 13–14

playing along, 4–6, 8, 11, 16, 19, 26, 42, 53, 113, 161, 194, 222–226
  methodological implications of, 6–7

popular music studies, 8

public/private, 13–14, 18–19, 32, 49–50, 105, 124, 182, 216, 224–225

race, 7, 222, 228n5, 229n9, 232n9
  in GTA, 26, 36, 38–39, 44–45, 46–51, 56, 62–63, 66, 70–71, 80–81, 228n13. *See also* blackface minstrelsy

rap. *See* hip-hop

RedOctane, 86–87

religion, 192

repertoire, 5, 7, 13, 15, 221–226
  and GTA, 11, 42, 44
  and *Guitar Hero* and *Rock Band*, 16, 87–89, 99, 105, 111–112, 119, 123, 126, 128, 140, 151, 231n28
  and online music pedagogy, 18, 157
  and online yoga pedagogy, 205, 209, 216

radio. *See* Grand Theft Auto (GTA)

repetition, 165, 189, 217, 228

Rock Band. See Guitar Hero and *Rock Band*

rock
  and authenticity, 14, 95, 99, 114, 127, 129
  and creativity, 14, 115, 151
  and gender, 95
  heroism, 16, 99, 103, 148, 151

and liveness, 14, 95, 98, 151
  performance, 120, 122–123, 125–126, 128–129, 149–150, 151. *See also Guitar Hero* and *Rock Band*

Rockstar Games, 24, 30, 35, 38–39, 42, 50, 57–59, 75

role-play, 43–44. *See also* massively multiplayer online game (MMOG)

Rose, Tricia, 79, 81

Sacred Harp singing, 137, 182, 221, 233n10

Samudra, Jaida Kim, 207, 218

Schafer, R. Murray, 12, 85–86

schizophonia, 12

schizophonic performance,
  definition of, 15, 85–86
  in *Guitar Hero* and *Rock Band*, 129, 142, 141–144, 149–151, 224

Schutz, Alfred, 5, 111

Scovil, Shaun, 108–109, 129, 132–135, 137. *See also Guitar Hero* and *Rock Band*

semiotic anxiety, 31

sensational knowledge, 5, 123, 141, 208, 209. *See also* Hahn, Tomie

Second Life, 7, 26, 28

sexuality,
  in GTA, 47, 51
  in *Guitar Hero* and *Rock Band*, 142–144, 223
  and Next Level Guitar, 171–173

Signifyin(g), 46–47, 79. *See also* Gates, Henry Louis, Jr.

social media, 183, 193, 209

somatic narratives, 218

sound-reproduction technology, 150, 164
  and amateur musical practice, 12
  and distinctions between live and recorded sound, 14, 86, 141–142

Sutton-Smith, Brian, 5, 10, 36, 49

symbolic creativity, 113

Taub, David, 18, 157–160, 162–163, 164–166, 167–171, 173–176, 178–181

technoculture, 19

techno-mediated practices, 14

theatricality, 125–129, 141–144, 149–151

thick description, 207
thick participation, 207, 216
Thompson, Jeff, 165, 168–169, 179
tourism
   and ethnography, 53
   and fieldwork, 32–37
   and GTA, 11, 28–32, 32–37, 52–53
transmission, 183, 221–222, 224. *See also* amateur-to-amateur (A2A); online music transmission; online music pedagogy

video game demographics, 6
virtual, 7–8, 150–151, 225, 227n4
   and visceral experience, 17, 158, 166–167, 182, 183, 193, 205
virtual performance, 5, 26–27, 161, 179, 225
virtuosity,
   in *Guitar Hero* and *Rock Band*, 16, 125, 142, 144–149

Waksman, Steve, 15, 114, 143
Web 2.0, 17, 217, 227n8
Wong, Freddie, 125–127, 129, 142–143, 144–149, 223, 232n10. *See also Guitar Hero* and *Rock Band*

yoga. *See* ashtanga yoga
YouTube, 4, 16–17, 223–225, 231n1
   and amateur-to-amateur transmission, 183, 184–186, 203, 206, 209, 210, 217
   and GTA, 27, 51, 54, 229n12
   and *Guitar Hero* and *Rock Band*, 16–17, 82, 94, 104, 116, 125–126, 139, 142–143, 146–147, 149
   and online music transmission, 156–158, 160, 162–163, 165, 170–171, 172–176, 180